Curriculum Issues in an Era of Common Core State Standards for Mathematics

Edited by

Christian R. Hirsch
Western Michigan University

Glenda T. Lappan
Michigan State University

Barbara J. Reys
University of Missouri–Columbia

Copyright © 2012 by
The National Council of Teachers of Mathematics, Inc.
1906 Association Drive, Reston, VA 20191-1502
(703) 620-9840; (800) 235-7566; www.nctm.org
All rights reserved

Library of Congress Cataloging-in-Publication Data

Curriculum issues in an era of common core state standards for mathematics / edited by Christian R. Hirsch, Glenda Lappan, Barbara J. Reys.
    p. cm.
  ISBN 978-0-87353-705-6
  1. Mathematics--Study and teaching—Cullicula—United States. 2. Curriculum planning—United States. 3. Teacher participation in curriculum planning—United States. I. Hirsch, Christian R. II. Lappan, Glenda. III. Reys, Barbara.
  QA13.C893 2012
  510.71'073—dc23

2012018284

The National Council of Teachers of Mathematics is a public voice of mathematics education, supporting teachers to ensure equitable mathematics learning of the highest quality for all students through vision, leadership, professional development, and research.

When forms, problems, and sample documents are included or are made available on NCTM's website, their use is authorized for educational purposes by educators and noncommercial or nonprofit entities that have purchased this book. Except for that use, permission to photocopy or use material electronically from *Curriculum Issues in an Era of Common Core State Standards for Mathematics,* must be obtained from www.copyright.com, or contact Copyright Clearance Center, Inc. (CCC), 222 Rosewood Drive, Danvers, MA 01923, 978-750-8400. CCC is a not-for-profit organization that provides licenses and registration for a variety of users. Permission does not automatically extend to any items identified as reprinted by permission of other publishers and copyright holders. Such items must be excluded unless separated permissions are obtained. It will be the responsibility of the user identify such materials and obtain the permissions.

The publications of the National Council of Teachers of Mathematics present a variety of viewpoints. The views expressed or implied in this publication, unless otherwise noted, should not be interpreted as official positions of the Council.

This book was prepared with support from the National Science Foundation (NSF) under grant no. ESI-0333879. Any opinions, findings, conclusions, or recommendations expressed in this volume are those of the authors and editors and do not necessarily reflect the views of the NSF.

Printed in the United States of America

Contents

Preface vii

How This Professional Development Resource Might Be Used ix

Section I: Introduction 1

    1. The Common Core State Standards for Mathematics: How Did We Get Here, and What Needs to Happen Next? 3
        Jere Confrey, North Carolina State University, Raleigh
        Erin E. Krupa, Montclair State University, Montclair, New Jersey

Section II: Interpreting and Responding to the Common Core State Standards for Mathematics 17

    Introduction 18

    Questions for Reflection and Collective Discussion 19

    2. Making the Transition to the Common Core State Standards for Mathematics 21
        Randall I. Charles, Emeritus, San Jose State University, California

    3. Common Core State Standards for Middle Grades Mathematics: Implications for Curriculum and Instruction 33
        Barbara J. Reys, University of Missouri–Columbia
        Barbara J. Dougherty, University of Missouri–Columbia
        Travis A. Olson, University of Nevada, Las Vegas
        Amanda Thomas, University of Missouri–Columbia

    4. Standards for High School Mathematics in the Common Core State Standards Era ... 47
        W. Gary Martin, Auburn University, Alabama
        Eric W. Hart, American University in Dubai, United Arab Emirates

Section III: Building Curriculum Coherence: Case Studies 61

    Introduction 62

    Questions for Reflection and Collective Discussion 62

    5. Using Curriculum to Build on Children's Thinking 65
        Corey Drake, Michigan State University, East Lansing
        Michelle Cirillo, University of Delaware, Newark
        Beth Herbel-Eisenmann, Michigan State University, East Lansing

6. Using Curriculum to Focus on Understanding..73
   Michelle Cirillo, University of Delaware, Newark
   Beth Herbel-Eisenmann, Michigan State University, East Lansing
   Corey Drake, Michigan State University, East Lansing

7. Adapting Curriculum to Focus on Authentic Mathematics...81
   Michelle Cirillo, University of Delaware, Newark
   Corey Drake, Michigan State University, East Lansing
   Beth Herbel-Eisenmann, Michigan State University, East Lansing

Section IV: Cultivating Mathematical Practices and Habits of Mind 89

Introduction..90

Questions for Reflection and Collective Discussion..91

8. An Algebraic-Habits-of-Mind Perspective on Elementary School Mathematics..........93
   E. Paul Goldenberg, Education Development Center, Newton, Massachusetts
   June Mark, Education Development Center, Newton, Massachusetts
   Al Cuoco, Education Development Center, Newton, Massachusetts

9. Developing Mathematical Habits of Mind..105
   June Mark, Education Development Center, Newton, Massachusetts
   Al Cuoco, Education Development Center, Newton, Massachusetts
   E. Paul Goldenberg, Education Development Center, Newton, Massachusetts
   Sarah Sword, Education Development Center, Newton, Massachusetts

10. Organizing a Curriculum around Mathematical Habits of Mind..............................111
    Al Cuoco, Education Development Center, Newton, Massachusetts
    E. Paul Goldenberg, Education Development Center, Newton, Massachusetts
    June Mark, Education Development Center, Newton, Massachusetts

Section V: Selecting and Strategically Using Technology Tools and Resources to Support Mathematics Teaching and Learning..121

Introduction..122

Questions for Reflection and Collective Discussion..122

11. Strategically Using Calculators in the Elementary Grades..125
    Kathryn B. Chval, University of Missouri–Columbia
    Sarah J. Hicks, Rockhurst University, Kansas City, Missouri

12. Technology and Mathematics in the Middle Grades 139
    Richard Hollenbeck, University of Maryland, College Park
    James Fey, University of Maryland, College Park

13. Using Computer Algebra Systems to Develop Big Ideas in Mathematics with Connections to the Common Core State Standards for Mathematics 149
    Rose Mary Zbiek, Pennsylvania State University, University Park
    M. Kathleen Heid, Pennsylvania State University, University Park

Section VI: Learning Progressions in School Mathematics: The Case of Statistics 161

Introduction 162

Questions for Reflection and Collective Discussion 162

14. Statistics in the Elementary Grades: Exploring Distributions of Data 165
    Christine A. Franklin, University of Georgia, Athens
    Denise A. Spangler, University of Georgia, Athens

15. Statistics in the Middle Grades: Understanding Center and Spread 175
    Gary Kader, Appalachian State University, Boone, North Carolina
    Jim Mamer, Rockway Middle School, Springfield, Ohio

16. Statistics in the High School Mathematics Curriculum: Building Sound Reasoning under Uncertain Conditions 185
    Richard Scheaffer, Emeritus, University of Florida, Gainesville
    Josh Tabor, Canyon del Oro High School, Tucson, Arizona

Section VII: Improving Vertical Articulation: Challenges and Promising Practices 195

Introduction 196

Questions for Reflection and Collective Discussion 196

17. Transitions from Elementary to Middle School Mathematics 199
    Janie Schielack, Texas A&M University, College Station
    Cathy L. Seeley, Charles A. Dana Center, University of Texas at Austin

18. Transitions from Middle School to High School: Crossing the Bridge 207
    Lisa C. Brown, Charles A. Dana Center, University of Texas at Austin
    Cathy L. Seeley, Charles A. Dana Center, University of Texas at Austin

19. High School to Postsecondary Education: Challenges of Transition 215
    Susan Hudson Hull, Charles A. Dana Center, University of Texas at Austin
    Cathy L. Seeley, Charles A. Dana Center, University of Texas at Austin

Preface

In March 1989, the National Council of Teachers of Mathematics' (NCTM) publication of *Curriculum and Evaluation Standards for School Mathematics* launched a two-decade effort to set standards at national and state levels as a means for improving teaching and learning in most school subjects. For school mathematics, these efforts culminated in June 2010 with the release of the *Common Core State Standards for Mathematics* (CCSSM), developed under the auspices of the National Governor's Association and the Council of Chief State School Officers (http://www.corestandards.org/). Forty-eight states collaborated in the development of the CCSSM; to date, 45 states, the District of Columbia, and the U.S. Virgin Islands have officially adopted the CCSSM to replace existing state standards.

The CCSSM presents new challenges and opportunities for local school districts and teachers to focus on curriculum as a means of improving what students understand and can do in their study of mathematics. Effective local implementation of the CCSSM will require careful district planning and attention to phase-in models and new norms of practice, including substantial conversations and planning across grades. It will require greater attention to important mathematical practices, curriculum coherence, and vertical articulation across grades K–12. It will also require mathematical knowledge for teaching new content expectations in ways that involve sense making and reasoning.

As a part of the work of the Center for the Study of Mathematics Curriculum (CSMC), our professional development efforts have focused on encouraging teachers to have conversations within and across grades. Teachers need to interact with one another, so that students see that the work they do in one grade connects to what they learn in the next grade and beyond. Through such interaction, teachers will develop expectations about what mathematics students will have experienced in the previous grade and what students will encounter in the next grade and beyond. Our students are wonderfully adept at emptying their minds over the summer, behaving in the next grade as if they have learned little. Our response is often to reteach rather than create opportunities to review while engaging in new material. Engagement across grades can help teachers diminish the loss of instructional time that reteaching causes.

CSMC leaders have developed this volume to further the goal of teachers having opportunities to interact across grades in ways that help both teachers and their students see connections in schooling as they progress through the grades. Each section of this volume contains three companion chapters appropriate to the three grade bands—K–5, 6–8, and 9–12—focusing on important curriculum issues related to understanding and implementing the CCSSM.

To promote a deeper understanding of CCSSM-related ideas and their implications for district policy and practice, an Introduction and a series of Questions for Reflection and Collective Discussion accompany each set of three grade-band chapters in Sections II–VII. The questions and the prompted collegial discussions, curriculum and related resource audits, CCSSM implementation planning, and reports of classroom trials of new ideas are the core of this volume. We envision groups of teachers working together, within and across grades in professional development settings, to accomplish this work. Such interactions around practice can help start conversations within and across buildings that change the culture and undescore the importance of teachers having time to learn, plan, and work together. We hope that you will find such interactions both educative and empowering as you work to interpret and implement the CCSSM effectively.

Acknowledgments

We are grateful to the National Science Foundation (grant no. ESI-0333879) for its support of the Center for the Study of Mathematics Curriculum and its multifaceted work, including the preparation of this volume. We thank the authors of the chapters for their responsiveness to our invitations and editorial suggestions and for the quality of their contributions.

We would like to thank Valerie Mills, Oakland County Schools, for reviewing the prospectus for this book and for her helpful comments and suggestions related to the book's professional development orientation.

We also wish to extend our gratitude to NCTM President J. Michael Shaughnessy for his encouragement and support.

Finally, we would like to gratefully acknowledge the superb assistance of Hope Smith at Western Michigan University in compiling, preparing, and reviewing the prepublication material.

How This Professional Development Resource Might Be Used

As the preface has indicated, we intend this volume as a professional development (PD) resource that educators can use flexibly with multiple audiences. Possible models are outlined below.

Model 1: District-Level Professional Development: A Summer Program with Follow-Up Sessions

District PD programs vary depending on the district's size and goals. One program could involve a sequence of four- or five-day sessions in the summer, focusing on understanding the Common Core State Standards for Mathematics (CCSSM) better. Before the first session, participants would download a personal copy of the CCSSM, available at http://www.corestandards.org/. They would review the document's structure—mathematical practices and content domains, clusters, and standards—and examine the standards related to their upcoming teaching assignment.

The summer sessions would focus on the implications of the CCSSM for the district, with the readings and Questions for Reflection and Collective Discussion in Sections I and II serving as primary resource materials. Grades K–5 teachers would be responsible for reading at least a section's first two chapters. Grades 6–8 teachers would read at least the first two chapters and scan the third, noting interesting or provocative ideas. Grades 9–12 teachers would read at least the second and third chapters and scan the first to help build a grades K–12 curriculum perspective.

Following the summer session, five follow-up, monthly, half- or full-day PD sessions would occur during the school year, one devoted to each of this volume's Sections III–VII. For these subsequent sessions, participants would preread and reflect on the appropriate section's assigned readings to prepare for the next PD session. Each session would give participants time to collaborate in grade levels or bands, discussing that section's selected questions. Each session should allow time for discussions across grades, reports on the groups' completed tasks, and summaries of implications for district planning and decision making. Note that in addition to completing readings prior to a session, participants would work on extended tasks (see, for example, page 64, question 14, and page 92, question 10) between or across sessions. This is especially true for sections that involve curriculum and instructional resource audits and readings from other sources accessible on the Web.

The school district can vary how this model organizes and groups PD participants on the basis of the district's size and goals.

Model 2: District-Level Professional Development: An Academic Year Program

This model is an academic-year variation on model 1. The school year would have seven to eight half- or full-day professional development sessions, scheduled at least a month apart. Again, before the first session, participants would download a copy of the CCSSM and review the document's structure and the standards related to their teaching assignment. This first session would focus on the structure and content of the CCSSM, with support materials from Section I. The rest of the PD program would unfold much like in model 1: participants would complete assigned readings in Sections II–VII before the appropriate session, where they would reflect on the readings and discuss related questions and implications for district policy and practice.

Model 3: Building-Level Professional Development

Building-level PD benefits teachers in many different configurations—by grades or grade bands, by course or course sequence at the high school level, in interested teachers' groups, and as a means of enculturating new teachers into the building and the profession. What makes such PD especially useful is that it has a focus both compelling and important to the success of teachers and their students. Since the CCSSM affects both the content expectations and mathematical practices for all students, the need to work together within a building to focus on the intended mathematics and its practices is especially compelling.

This book provides both information and materials for PD that engages in productive, within-building collaboration and planning. The collection of chapters includes readings appropriate for elementary, middle, and high school study or planning groups. Although all the chapters can help one understand grades K–12 teaching challenges, grade and grade-band focuses can help build lasting collaborations more effectively within a building that enhance mathematics teaching and learning. This means that teachers talk across and within grades.

A group within a school building can make an excellent start by reading the Section I chapter by Confrey and Krupa, on CCSSM background and strategic next steps. Section II focuses on interpreting and responding to the CCSSM at different grade bands. The group can use the accompanying questions to discuss the chapter(s). A group of elementary school teachers can read the Charles chapter and discuss its ideas relative to their classroom practices. Middle school teachers can read the Reys et al. chapter and discuss how they can help their students succeed in meeting the CCSSM expectations for that grade band. High school teachers can focus on the Martin and Hart chapter to help develop their collective approaches to engaging their students in mathematics in ways that both make sense and help reach the CCSSM's mathematical expectations.

Reading and discussing the chapters in Section III can help teacher groups focus more directly on children's thinking and on looking for evidence that their students understand the mathematics they encounter. Chapters in Section IV focus on developing mathematical habits of mind. Reading and discussing these chapters can help teachers examine their own mathematical habits of mind. Discussing individual ways of approaching mathematical situations can result in a deeper understanding of what mathematical practices would be productive to promote with children at their various grade bands or grades.

Technology has become an aid to developing mathematical content, exploring mathematical situations, and problem solving. Chapters in Section V can help participants look more deeply into how technology can both become a learning tool *and* support and enhance students' progress in mathematics and in developing mathematical thinking and reasoning.

Statistics is pervasive in contemporary society and has become an important part of the mathematics curriculum at all grade levels. Statistical explorations engage students at all levels. Discussion of chapters in Section VI can help participants build grade-appropriate statistical activities for their students. The chapters' activities highlight the usefulness of statistical reasoning and its importance in our lives. Discussions across grade bands can lead to productive consideration of learning progressions, both in statistics and, by analogy, in other mathematical domains.

For teachers within a school building, exploring this volume's final set of chapters, on vertical articulation, is especially important. Transitions between grades challenge students and teachers. Teachers need opportunities to talk across grades so that mathematics learning across grades remains seamless. Focusing on these chapters can help teachers create and maintain that seamlessness. Discussion across grades has yet another advantage for teachers: it improves their collective understanding of the important mathematics that students are studying and how that mathematics builds across a span of grades.

Model 4: A Master's-Degree Course for In-Service Teachers

Using *Curriculum Issues in an Era of Common Core State Standards* can help engage in-service teachers in considering how the CCSSM affects their school curriculum materials and instructional practices. For example, after introducing and reviewing the structure of the CCSSM, as in the case of the other PD models, this model can use each section of the book as the basis for assigned readings and class discussion (e.g., one section per session or week). The choice of which chapters in a section to assign as pre-reading material would depend on the class's makeup. For example, if the class includes only elementary school teachers, the course could assign and discuss the elementary and middle school chapters. If the class includes only middle school teachers, then they all should read the middle school chapters, be prepared to discuss them, and complete questions and activities related to them. Half the teachers in the class should also read and be prepared to report on the elementary school chapters; the other half would have similar responsibility for the high school chapters. Finally, for a class consisting solely of high school teachers, the course can assign the middle and high school chapters.

Completing such a course would position participants well to take on leadership roles in planning and monitoring CCSSM implementation in their school and district.

In a broader curriculum course, the instructor can select a subset of this book's sections for use. We recommend that all groups use Sections I and II as introductory material, and that they all make Section IV, mathematical practices, a priority.

For additional CCSSM-related resources usable in a master's course for in-service teachers, see the Center for the Study of Mathematics Curriculum website, http://www.mathcurriculumcenter.org.

Section I
Introduction

Chapter 1

The Common Core State Standards for Mathematics: How Did We Get Here, and What Needs to Happen Next?

Jere Confrey
Erin E. Krupa

ETYMOLOGICALLY, the word *standard* comes from the Anglo-French word *estaundart*, referring to a flag displayed on a battlefield to rally the troops (Oxford English Dictionary n.d.). Over time, the term evolved in two ways. First, instead of referring to a king's authority, it came to mean a consensus among experts. Second, it evolved to mean improved technical specifications that promote efficiency and make measures of that efficiency easier. Standards in education serve two similar purposes: they express a consensus among experts of what to teach and when to teach it, and they make measuring students' proficiency easier through assessments.

For the first time in U.S. history, states across the nation are rallying behind a common set of voluntary state mathematics standards. Spirited debate among varied experts—mathematicians, mathematics educators, teachers, statisticians, and policy leaders—has been undertaken to reach consensus on a negotiated set of goal statements and to build common assessments. Led by the National Governors Association (NGA) and the Council of Chief State School Officers (CCSSO), 45 states, the District of Columbia, and the U.S. Virgin Islands have adopted the Common Core State Standards for Mathematics (CCSSM), released June 2, 2010.

These states, the District of Columbia, and the U.S. Virgin Islands have an historic opportunity to ensure that school mathematics programs are based on "fewer, clearer, and higher" standards (Common Core State Standards Initiative [CCSSI] 2010, p. 1) designed to prepare students with the knowledge and skills they need to succeed in college and work.

This chapter traces content standards' evolution in mathematics education, summarizes the CCSSM's current state, and discusses the next steps to ensure the proper implementation of the CCSSM. Typically mathematics educators have defined *content*

This chapter is based on two papers written for the Center for the Study of Mathematics Curriculum, Confrey (2007) and Confrey and Krupa (2010).

standards as a specification of "what a person should know or be able to do" (National Research Council 2002, p. 2), and have stated that standards should "indicate the topics and skills that should be taught at various grades or grade spans and are intended to guide public school instruction, curriculum, teacher preparation, and assessment" (Goertz 2010, p. 52). We use a revised definition reflecting the evolution of standards (Confrey 2007, pp. 6–7):

> Content standards consist of a negotiated settlement among authorized experts concerning the specification of what a person should know or be able to do, with consideration of how that is to be measured and/or documented, and as a means of modulating or effecting change within the system of education and restricting excessive variation.

Both definitions recognize the importance of standards functioning collaboratively with other elements in the system—curriculum, instructional teaching capacity, professional development, and assessment.

Part 1: The History of Standards in Mathematics Education

The U. S. Constitution prohibits federal agencies from regulating content standards and assessments, leaving states in charge of their own educational system (Fuhrman 2004). Attempts to standardize education date back to the Committee of Ten in 1892, which sought to standardize secondary school curriculum in order to prepare college-bound and workforce-ready students. Ten years after the committee's inception, Dexter (1906) questioned its influence, wondering how closely the changes in high school curriculum coincided with the committee's recommendations. He concluded that the report had little influence on the pedagogy and content taught in secondary school classrooms, raising what has become a perennial question of how to influence education in the classroom.

When the Soviet Union launched the *Sputnik* satellite in 1957, the United States began another concerted effort to produce content standards, focusing on "high quality mathematics for college-capable students, particularly those heading for technical or scientific careers" (National Advisory Committee on Mathematics Education 1975). These standards produced curricula dominated by formal structures, such as set theory and deductive proof. Then in 1975, the Conference Board of the Mathematical Sciences reexamined these ideas and recommended that content standards should (*a*) maintain the emphasis on the logical structure of mathematics, (*b*) integrate concrete experiences, (*c*) include applications, and (*d*) foster the use of symbols and formal notation. They further advised that mathematics instruction include (*a*) calculators by eighth grade, (*b*) the metric system, and (*c*) statistics (National Advisory Committee on Mathematics Education 1975, pp. 136–39).

The following paragraph, taken from Confrey (2007, pp. 13–14), summarizes the state of mathematics standards at the end of the twentieth century.

> In 1989, after three years of work led by Thomas Romberg, together with mathematics teachers, researchers, and administrators, the National Council of Teachers of Mathematics produced the *NCTM Curriculum and Evaluation Standards*, with a set of *Professional Standards for Teaching Mathematics* (1991) and *Assessment Standards for School Mathematics* (1995). Only the *Curriculum*

and Evaluation Standards exerted considerable influence on other factors in reform, curriculum, state changes, [and] assessment. The standards were intended to ensure quality, identify explicit goals, and promote change. They were purposed to create mathematically literate workers, encourage lifelong learning, provide opportunities for all, and support an informed electorate. They were structured by grade bands (K–4, 5–8, and 9–12), and each addressed standards of problem solving, communication, reasoning, connections, and estimation. They addressed the content strands of number and numeration, geometry, measurement, statistics and probability, algebra and trigonometry, and discrete mathematics. Issues of pedagogy were integrated into issues of content, emphasizing the importance of active participation in learning by students. The standards drew heavily on research on student thinking, student misconceptions, and how students learned particular ideas as they encountered challenging tasks. They warned against relying too heavily on memorization and procedural understanding, based on numerous studies documenting disintegration of students' apparent knowledge when asked for reasons and explanations, and stressed conceptual understanding. (Erlwanger 1973; Ginsburg 1991; Kamii 1985)

In response to the development of standards in mathematics and science, the federal government chose to use standards as a means to establish a more rigorous educational system, known as "the standards movement" or "systemic reform" (Cohen 1995; Lewis 1995). As noted by Goertz (2010, p. 54):

The Improving America's Schools Act of 1994 required states to develop challenging content standards in at least reading and mathematics, create high-quality assessments to measure performance against these standards, and have local districts identify low-performing schools for assistance.... With the enactment of the NCLB Act of 2001, the federal government expanded its role significantly, requiring states to test more frequently and set more ambitious and uniform improvement goals for their schools, and prescribing sanctions for schools that fail to meet these goals.

Increased accountability pressured states to align their standards with ones produced by national organizations. Nonetheless, the mathematics frameworks in the 50 states and the District of Columbia still varied significantly (Reys 2006). The first ten years of the twenty-first century saw numerous standards released, from prominent national organizations, to help states develop more coherent frameworks. NCTM published *Principles and Standards for School Mathematics* (2000), *Curriculum Focal Points* (2006), and *Focus in High School Mathematics: Reasoning and Sense Making* (2009); Achieve, *Mathematics Benchmarks: Grades K–12* (2004); the American Statistical Association, *Guidelines for Assessment and Instruction in Statistics Education (GAISE)* (Franklin et al. 2007); the College Board (2006), "College Board Standards for College Success: Mathematics and Statistics"; and the National Mathematics Advisory Panel (2008), *Foundations for Success*.

Among these standards-setting efforts a consensus was emerging on the need for specificity in grades and courses, and that these specifications need to be "important and challenging" for all students (Confrey 2007). Usiskin (2010, p. 34) makes the following point:

[These standards] brought algebra and some data analysis into the elementary school, algebra into grade 8 and earlier for many students, applications into the algebra and geometry curricula, graphing calculators into the study of functions, and a major increase in the number of high school students taking calculus. It promoted active learning, classroom discourse, alternative algorithms, and multiple ways of approaching problems.

The progress made in creating and refining national standards for mathematics is undeniable. In less than twenty years, we have progressed as a nation from no standards, to multiple sets of standards, to a voluntary set (CCSSM), each created with clear attention to salient features affecting students' learning. In a distributed system of education, we have crafted a means to advise diverse constituencies on what students should know and do. We have seen many of the states—the national unit of educational change—adopt or adapt this means for local consumption. We have also clearly embraced the complexity involved in acknowledging diverse groups of experts and their crucial roles in establishing standards.

Part 2: The Current State of the CCSSM in Mathematics Education

The development of the CCSSM began in July 2009. The CCSSI based the document's content on benchmarking of U.S. standards to high-performing countries on international exams, research on students' mathematical knowledge and skill from the National Mathematics Advisory Panel, and research from students on student learning (CCSSI 2010, p. 1). The lead writers valued information on how students learn mathematics and, as Confrey (2007, p. 33) stated, the development of "sequenced obstacles and challenges for students ... absent the insights about meaning that derive from careful study of learning, would be unfortunate and unwise." The writers also had to weigh competing sources of research and decide what to emphasize and what to delete. They chose to separate content and process standards and did so by proposing eight mathematical practices. These eight practices, derived from a combination of NCTM's process standards and mathematical proficiencies from the National Research Council's (2001) report *Adding It Up*, serve as a foundation for grades K–12 mathematics instruction:

1. Make sense of problems and persevere in solving them
2. Reason abstractly and quantitatively
3. Construct viable arguments and critique the reasoning of others
4. Model with mathematics
5. Use appropriate tools strategically
6. Attend to precision
7. Look for and make use of structure
8. Look for and express regularity in repeated reasoning

These eight practices offer instructional habits of mind for teachers to use as they teach students how to understand the content standards. Confrey and Krupa (2010, p. 9) emphasized the importance of linking content and practice standards as a means to strengthen understanding, arguing that "the practices sustain mathematics as the content evolves. As such, they make what students learn enduring and they ensure that students will continue to be prepared to learn new mathematics."

Figure 1.1 depicts the content standards' design and organization. The *standards* contain the first component of Confrey's (2007) definition, "what a person should know or be able to do." The document groups these standards into *clusters* of related standards, which it further groups into larger *domains* of related clusters.

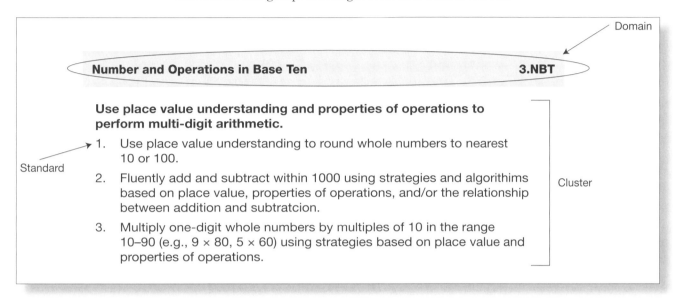

Fig. 1.1. An example of a content standard, taken from the CCSSM's introduction (CCSSI 2010, p. 5)

For high school, the CCSSM clusters the standards into conceptual categories: Number and Quantity, Algebra, Functions, Modeling, Geometry, and Statistics and Probability. These clusters contain the mathematical content all students should learn for college and career readiness, but the CCSSM does not mandate the order in which the content should be taught.

The CCSSI designed these content standards to support learning trajectories, which describe the transition in students' thinking from novice to sophisticated views of "big ideas" on the basis of empirical research, and identify common obstacles and landmarks in that process. Learning trajectories recognize the pivotal roles that instruction, careful sequencing and selecting of curricular tasks, and proper use of tools and language to support learning all play. Not all "big ideas" have equally solid research bases. Further, some decisions by the lead writers represent "logical thought experiments" on how trajectories might proceed: readers should view these as conjectures subject to further research. As the CCSSM states (CCSSI 2010, p. 5), "one promise of common state standards is that over time they will allow research on learning progressions to inform and improve the design of standards to a much greater extent than is possible today." Learning trajectories are important because they allow teachers to know what to expect about entering students' preparation, what to teach in recognition of what they will expect of their students, and how to relate concepts across strands at each grade. Teachers report that these help them recognize a variety of students' ideas, support discourse, and engage in rich uses of classroom assessment. Confrey, Maloney, and Nguyen (2011) have created a "learning trajectories display" of the CCSSM, which shows how concepts evolve over time across grades (K–8) and difficulty levels (high school). Because of the two-dimensional structure at high school, the learning trajectory display supports either a siloed or an integrated curriculum, remaining agnostic about the curriculum or order in which it delivers content.

Table 1.1 details several major shifts in the CCSSM in grades K–5 for most states. These grades K–5 standards emphasize numeration and operation, along with many concepts and skills introduced at earlier grade levels. They also expect students to learn three systems of measurement simultaneously—nonstandard, English, and metric. An increased emphasis exists on fractions as numbers and on using the number line to give students structure and visualization of the number system. These standards have deemphasized or removed early algebra—as introduced through patterns; statistics and probability; and percent, ratio, and proportion—from grades K–5. The writers' arguments were (*a*) having fewer topics means some must be eliminated, and (*b*) emphasizing number mastery would lead to quicker and more secure learning of these eliminated topics in middle grades. Many educators question these assumptions and either will argue for their inclusion in future revisions on the basis of empirical studies or choose to address the topics in the 15 percent discretionary content left up to states.

Table 1.1
The CCSSM Grades K–5 Domains

Domains	Grades
Counting and Cardinality	K only
Operations and Algebraic Thinking	K–5
Number and Operations in Base Ten	K–5
Number and Operations—Fractions	3–5
Measurement and Data	K–5
Geometry	K–5

The most extensive changes in the CCSSM occur at the middle grades (see table 1.2), beginning with a strong foundation for early algebra in ratio and rate, and with an introduction to statistics. Grade 6 alone houses the beginning of ratio, proportion, percents, and statistics. Grade 7 introduces rational number. One-third of the regular grade 8 curriculum consists of algebra topics for all students. Phil Daro, one of the writers of the CCSSM, said, "The CCSS were written to assume 100% mastery, in any given year, of the preceding year's standards" (Confrey and Krupa 2010, p. 2). Thus, middle grades teachers should anticipate that, by having fewer elementary school standards, the richness of the grades K–5 standards will produce more students who understand the concepts and will result in less need for remediation and repetition.

Table 1.2
The CCSSM Grades 6–8 Domains

Domains	Grades
Ratio and Proportional Relationship	6–7
The Number System	6–8
Expressions and Equations	6–8
Functions	8
Geometry	6–8
Statistics and Probability	6–8

Table 1.3 presents the standards for grades 9–12 by conceptual categories that schools must address over the four years a student is in high school, without referent to a particular curricular approach. The major changes at the high school level are the masteries students must show beyond traditional Algebra 2 content. These topics include periodic functions, polynomials, radicals, advanced probability and statistics, and mathematical modeling. In table 1.3, note the absence of any topics in the domains column for the "modeling" conceptual category. The writers of the CCSSM did not incorporate a specific modeling standard, but instead suggested that modeling be woven throughout other standards.

Table 1.3
The CCSSM Grades 9–12 Conceptual Categories and Domains

Conceptual Categories	Domains
Number and Quantity	The Real Number System, Quantities, The Complex Number System, Vector and Matrix Quantities
Algebra	Seeing Structure in Expressions, Arithmetic with Polynomials and Rational Expressions, Creating Equations, Reasoning with Equations and Inequalities
Functions Overview	Interpreting Functions; Building Functions; Linear, Quadratic, and Exponential Models; Trigonometric Functions
Modeling	
Geometry	Congruence; Similarity, Right Triangles, and Trigonometry; Circles; Expressing Geometric Properties with Equations; Geometric Measurement and Dimension; Modeling with Geometry
Statistics and Probability	Interpreting Categorical and Quantitative Data, Making Inferences and Justifying Conclusions, Conditional Probability and the Rules of Probability, Using Probability to Make Decisions

The CCSSM is leading to some significant activity in developing new assessments, an approach that builds on the view of standards as a means to consider "how that [what students should know] is to be measured and/or documented" (Confrey 2007, pp. 6–7). The U. S. Department of Education is funding two state-based assessment consortia, the Partnership for Assessment of Readiness for College and Career (PARCC) and the Smarter Balanced Assessment Consortium (SBAC), to align assessments to the CCSSM by the 2014–15 school year. Having these two groups create, test, validate, and disseminate innovative assessment systems in less than four years is an ambitious goal; collaborations across states, however, will be more efficient and should foster greater coherence than the current accountability system.

Finally, the third component of Confrey's (2007) definition of *standard* describes standards' role in "restricting excessive variation." Adopting the Common Core State Standards (CCSS) means that states agree to implement, word-for-word, 100 percent of the English language arts and mathematics standards. Each state, however, has the option to incorporate up to 15 percent self-selected standards in addition to the CCSS. According to the criteria used to write the CCSS (CCSSI 2010), adherence to these basic standards ensures coherence across states to assist with students' mobility and provides rigorous content and application for global competition. In addition, the CCSSM's writers have benchmarked their standards internationally and built them on the strengths of current state standards to give students the knowledge required for success in the twenty-first century. Although the standards' implementation will be key, agreement by states to follow the CCSS word-for-word is a first step in decreasing variation among teachers, schools, districts, and states. Mathematics teachers should also review the English language arts standards listed under "Scientific and Technical Reading and Writing." These standards, which lend the mathematical practices standards considerable support, express the importance of all students being able to apply mathematics, scientific and logical reasoning, and argumentation.

Part 3: Current and Future Recommendations

"While the development of the CCSS and widespread adoption is an accomplishment, in fact its passage marks only the beginning of the work to be done through professional development, creation of instructional materials and related tools, and phased implementation" (Confrey and Krupa 2010, p. 2). States will have numerous items to consider as they begin to incorporate the CCSS into instruction. Two frameworks can help in understanding the complexities involved in the implementation of the standards. First, the National Research Council's (2002) *Investigating the Influence of Standards: A Framework for Research in Mathematics, Science, and Technology Education* provides a comprehensive display of forces and channels that need examination in order to achieve the desired impact on students' learning (see fig. 1.2).

Second, Confrey and Maloney's (in press) framework presents standards and high-stakes tests as the bookends of the instructional process. The framework connects the two to show that the data from the high-stakes exams guide both instructional practice and standards' development. In figure 1.3, the triangle indicates that effective classroom practice, the heart of quality instruction, operates through the interactions among implemented curriculum, instructional practices, and forms of classroom assessment. The

How has the system responded to the introduction of nationally developed standards?			**What are the consequences for student learning?**

Contextual Forces
- Politicians and Policy Makers
- Public
- Business and Industry
- Professional Organizations

Channels of Influence Within the Education System

Curriculum
- State, district policy decisions
- Instructional materials development
- Text, materials selection

Teacher Development
- Initial preparation
- Certification
- Professional development

Assessment and Accountability
- Accountability systems
- Classroom assessment
- State, district assessment
- College entrance, placement practices

Teachers and Teaching Practice in classroom and school contexts

Among teachers who have been exposed to nationally developed standards—
- How have they received and interpreted those standards?
- What actions have they taken in response?
- What, if anything, about their classroom practice has changed?
- Who has been affected and how?

Student Learning

Among students who have been exposed to standards-based practice—
- How have students learning and achievement changed?
- Who has been affected and how?

Within the education system and in its context—
- How are nationally developed standards being received and interpreted?
- What actions have been taken in response?
- What has changed as a result?
- What components of the system have been affected and how?

Fig. 1.2. Factors to consider when implementing standards (NRC 2002)

double arrow to Professional Development signals that fostering rich classroom practices is enhanced when professional development is provided by teachers and when "best practices" of experienced teachers, along with research, guide the professional development.

The second framework complements the first with its focus on the effects of the CCSS on classroom practice in order to effect student outcomes. Although each framework centers on the instructional core, each framework's outcome is on students' learning, documented through assessment. The rest of this section will focus on our recommendations for what should be done now to implement the CCSSM properly, followed by future considerations to keep the current momentum moving in a positive direction.

The most crucial initial steps, to be taken now to ensure the proper, widespread implementation of the CCSSM, are considering phasing models and launching professional development. Compared to most of the current state standards, the CCSSM contains significant modifications of topics taught at each grade level; thus, local and state mathematics education communities should take appropriate phasing models into account. These should begin with awareness, by all stakeholders, and proceed toward specific guidelines for coordinated implementation across grades (Confrey and Krupa 2010), concentrating on the trajectories. We consider important stakeholders to include district-level administrators, principals, coaches, teachers, parents, university faculty in mathematics and mathematics education, external professional development providers, and students. The phasing model discussions should include strategies for the transition grades, 5–6 and 8–9, where changes in schools and schooling practices necessitate careful attention to students' success.

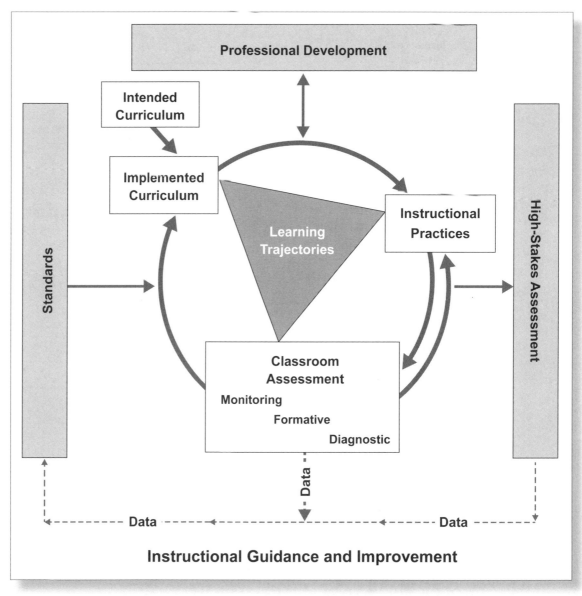

Fig. 1.3. Embedding the instructional core in an accountability framework (Confrey and Maloney in press)

Adequate professional development is imperative to ensure teachers are knowledgeable not only about the major shifts in the CCSSM, but more importantly, so that teachers can begin to lead with the mathematical practices. Since teachers typically use the same instructional practices used by their teachers (Ball 1988; Tyack and Cuban 1995), quality professional development should be designed and disseminated to help teachers negotiate the eight mathematical practices in relation to the content standards. This will help to improve and strengthen teachers' mathematical content and pedagogical knowledge (Loucks-Horsley et al. 2003).

Long-term goals, to be immediately convened, include the development of new summative assessments (under the direction of the two consortia), the development of effective formative and diagnostic assessment systems, the selection and revisions of curricular materials aligned with the content and practices of the CCSSM, and the creation of a process to ensure that the CCSSM remains a living document (Confrey and Krupa 2010; Krupa 2011). Work at the national level has already begun on the summative assessments; as this work continues, sharing insights and expectations with teachers will be important, so that they can incorporate assessments seamlessly into the 2014–15 school year.

To strengthen the instructional guidance teachers provide to each student, formative and diagnostic assessment systems should be embedded into teachers' instruction. These should be designed so that student understanding of main topics is apparent, providing teachers with the information they need to take the next steps with their instruction. These should also be designed to help teachers think critically about students' thinking and problem-solving strategies. Together these systems will help determine which steps need to be taken, each year, for a student to move through a successful learning progression. For more information on specific steps to be taken to create a formative and diagnostic assessment system, see Confrey (2011).

Curricula designers and publishers, front-line implementers from states and school districts, policy experts, and mathematics education researchers all share a fundamental assumption—that *curricula matter!* The scale of the change that the CCSSM proposes requires designated curricular materials that align with the content standards and mathematical practices (Confrey and Krupa 2010). Since the mathematical practices are presented independent from the content standards, there is a risk that the practices will be underemphasized. Curriculum developers need to align current materials to the CCSSM to produce high-quality mathematics curricula. Textbook adoption committees need to pay careful attention to distinguishing between materials that genuinely align to the CCSSM and those that claim alignment as a marketing strategy. Further, both communities need to ensure that their product or textbook selection embeds the mathematical practices throughout.

Finally, and most important, to guarantee the CCSSM's long-term success, and consequently students' understanding of mathematics, the CCSSM should remain a living document. Incorporating all the components of implementation depicted in figures 1.2 and 1.3, the entire mathematics education community should dedicate themselves to the standards' continual improvement. This improvement should include short-term fixes, medium-term adjustments, and long-term review and modification (Confrey and Krupa 2010). Revising the standards over time, on the basis of evidence collected from the field, is imperative. The success of the CCSSM will rest on whether they increase equity in outcomes, because clear delineation of curricular targets and better forms of testing should permit more effective instruction delivery. In addition, implementing the CCSSM should improve the United States' relative performance on international assessments.

Part 4: Conclusions

This is an exciting time for the future of mathematics education, but one that will require experts to collaborate to accomplish our shared goal of strengthening students' learning. The history of mathematics standards development in the United States suggests that even in a view recognizing states' rights, a coordinated effort to improve mathematics education should serve our highly mobile population better. Reviewing the CCSSM content suggests that needs will exist for teachers' professional development and for using communities of practice to converse and collaborate on plans for implementation. Furthermore, the position outlined in this chapter argues for the importance of teachers' attention to practices to ensure students not only learn the content specified at grade level but have adequate preparation to succeed at more advanced levels because of their facility with the practices. It also argues that the CCSSM must be a living document, owned by our professional community, to ensure that future revisions reflect what we learn from the standards' implementation process.

References

Achieve, Inc. *Mathematics Benchmarks: Grades K–12*. Washington, D.C.: Achieve, 2004.

Ball, Deborah L. "Unlearning to Teach Mathematics." *For the Learning of Mathematics* 8, no. 1 (1988): 40–48.

Cohen, David K. "What Is the System in Systemic Reform?" *Educational Researcher* 24, no. 9 (December 1995): 11–31.

College Board. "College Board Standards for College Success: Mathematics and Statistics." 2006. http://www.collegeboard.com/prod_downloads/about/association/academic/mathematics-statistics_cbscs.pdf.

Common Core State Standards Initiative [CCSSI]. *Common Core State Standards for Mathematics*. Washington, D.C.: National Governors Association Center for Best Practices and the Council of Chief State School Officers, 2010. http://www.corestandards.org.

Confrey, Jere. "Tracing the Evolution of Mathematics Content Standards in the United States: Looking Back and Projecting Forward toward National Standards." Paper presented at the Conference on K–12 Mathematics Curriculum Standards, Arlington, Va., 2007.

―――. *A Summary Report from the Conference Designing Technology-Enabled Diagnostic Assessments for K–12 Mathematics*. Friday Institute for Educational Innovation and North Carolina State University College of Education. 2011.

Confrey, Jere, and Erin E. Krupa. *Curriculum Design, Development, and Implementation in an Era of Common Core State Standards: Summary Report of a Conference*. Arlington, Va.: Center for the Study of Mathematics Curriculum, 2010.

Confrey, Jere, and Alan P. Maloney. "Next Generation Digital Classroom Assessment Based on Learning Trajectories in Mathematics." In *Steps toward a Digital Teaching Platform*, edited by Chris Dede and John Richards, in press.

Confrey, Jere, Alan P. Maloney, and Kenny H. Nguyen. *Learning Trajectory Display of the Common Core State Standards for Mathematics, Grades K–5, 6–8, and High School*. New York: Wireless Generation, 2011.

Dexter, Edwin Grant. "Ten Years' Influence of the Report of the Committee of Ten." *School Review* 14, no. 4 (1906): 254–69.

Erlwanger, Stanley H. "Benny's Concepts of Rules and Answers in IPI Mathematics." *Journal of Children's Mathematical Behavior* 1 (1973): 7–26.

Franklin, Christine, Gary Kader, Denise Mewborn, Jerry Moreno, Roxy Peck, Mike Perry, and Richard Scheaffer. *Guidelines for Assessment and Instruction in Statistics Education (GAISE) Report: A Pre-K–12 Curriculum Framework*. Alexandria, Va.: American Statistical Association, 2007.

Fuhrman, Susan H. "Less than Meets the Eye: Standards, Testing, and Fear of Federal Control." In *Who's in Charge Here? The Tangled Web of School Governance and Policy*, edited by Noel Epstein, pp. 131–63. Washington, D.C.: Brookings Institution Press, 2004.

Ginsburg, Herbert P. *Children's Arithmetic*. Austin, Tex.: Pro-Ed, 1991.

Goertz, Margaret E. "National Standards: Lessons from the Past, Directions for the Future." In *Mathematics Curriculum: Issues, Trends, and Future Directions*, Seventy-second Yearbook of the National Council of Teachers of Mathematics (NCTM), edited by Barbara J. Reys and Robert E. Reys, pp. 51–64. Reston, Va.: NCTM, 2010.

Kamii, Constance K. *Young Children Reinvent Arithmetic: Implications of Piaget's Theory*. New York: Columbia University, Teachers College Press, 1985.

Krupa, Erin E. *Curriculum and Assessment and the Common Core State Standards for Mathematics: Summary Report of a Conference*. Arlington, Va.: Center for the Study of Mathematics Curriculum, 2011.

Lewis, Anne C. "An Overview of the Standards Movement." *Phi Delta Kappan* 76, no. 10 (June 1995): 744–50.

Loucks-Horsley, Susan, Nancy B. Love, Katherine E. Stiles, Susan E. Mundry, and Peter W. Hewson. *Designing Professional Development for Teachers of Science and Mathematics*. Thousand Oaks, Calif.: Corwin Press, 2003.

National Advisory Committee on Mathematics Education. *Overview and Analysis of School Mathematics Grades K–12: The NACOME Report*. Washington, D.C.: Conference Board of the Mathematical Sciences, 1975.

National Council of Teachers of Mathematics (NCTM). *Curriculum and Evaluation Standards for School Mathematics*. Reston, Va.: NCTM, 1989.

———. *Professional Standards for Teaching Mathematics*. Reston, Va.: NCTM, 1991.

———. *Assessment Standards for School Mathematics*. Reston, Va.: NCTM, 1995.

———. *Principles and Standards for School Mathematics*. Reston, Va.: NCTM, 2000.

———. *Curriculum Focal Points for Prekindergarten through Grade 8 Mathematics: A Quest for Coherence*. Reston, Va.: NCTM, 2006.

———. *Focus in High School Mathematics: Reasoning and Sense Making*. Reston, Va.: NCTM, 2009.

National Mathematics Advisory Panel. *Foundations for Success: The Final Report of the National Mathematics Advisory Panel*. Washington, D.C.: U.S. Department of Education, 2008.

National Research Council. *Adding It Up: Helping Children Learn Mathematics*: Washington, D.C.: National Academy Press, 2001.

———. *Investigating the Influence of Standards*. Washington, D.C.: National Academy Press, 2002.

Oxford English Dictionary. *Oxford English Dictionary Standard*, n.d. http://www. dictionary.oed.com.www.lib.ncsu.edu:2048/cgi/entry/50236067?query_type=word&queryword=standard&first=1&max_to_show=10&sort_type=alpha&result_place=1&search_id=w7RD-5r5Jwz-18817&hilite=50236067.

Reys, Barbara J., ed. *The Intended Mathematics Curriculum as Represented in State-Level Curriculum Standards: Consensus or Confusion?* Charlotte, N.C.: Information Age Publishing, Inc., 2006.

Tyack, David, and Larry Cuban. *Tinkering toward Utopia: A Century of Public School Reform*. Cambridge, Mass.: Harvard University Press, 1995.

Usiskin, Zalman. "The Current State of the School Mathematics Curriculum." In *Mathematics Curriculum: Issues, Trends, and Future Directions*, Seventy-second Yearbook of the National Council of Teachers of Mathematics (NCTM), edited by Barbara J. Reys and Robert E. Reys, pp . 25–40. Reston, Va.: NCTM, 2010.

Section II
Interpreting and Responding to the Common Core State Standards for Mathematics

Introduction

NCTM's vision for school mathematics includes a call for curriculum that is "coherent, focused on important mathematics, and well articulated across the grades" (NCTM 2000, p. 14). A pivotal school improvement strategy over the past two decades has focused on establishing curriculum standards and then monitoring students' attainment of those standards through high-stakes assessments. The No Child Left Behind Act's passage adopted this strategy at the federal level, requiring that states adopt "challenging academic content standards" in mathematics, reading and language arts, and science that (*a*) specify what students are expected to know and be able to do, (*b*) contain coherent and rigorous content, and (*c*) encourage teaching advanced skills in these areas.

Although not obligatory, NCTM's *Standards* documents (1989, 2000) influenced the general content of states' and districts' curriculum standards as well as the discourse about good teaching and assessment practices. Whereas the NCTM *Standards* served as a general template, state standards were more specific due, in part, to the need to specify, for assessment purposes, the mathematics that a state expected all its students to learn in particular grades or courses.

The Center for the Study of Mathematics Curriculum's review of states' mathematics curriculum standards (Reys 2006) confirmed that statements of learning goals in state standards varied along several dimensions, including grain size (e.g., specificity), language used to convey learning goals (e.g., understand, explore, memorize, and so on), and the grade placement of specific learning goals. The variability in grade placement of many mathematical topics made it difficult for textbook publishers to create focused, coherent curriculum materials. This led state governors to agree in March 2009 to collaborate on developing common standards in mathematics and English/language arts. The goal was to increase the quality and rigor of mathematics standards and to establish consensus on the most important goals of school mathematics. The National Governors Association Center for Best Practices (NGA Center) and the Council of Chief State School Officers (CCSSO) assembled a small writing group to produce a set of common core standards for mathematics and English/language arts. In June 2010, fifteen months after the governors' agreement, the NGA Center and CCSSO released the *Common Core State Standards for Mathematics* (CCSSM; see http://www.corestandards.org/).

The adoption of the CCSSM by all but a few states (Alaska, Minnesota, Nebraska, Texas, Virginia) represents a revolution. For the first time in the United States, a large majority of schools, teachers, and students will focus on common, specific, and, at grades K–8, grade-level-focused learning goals for mathematics. Coupled with state collaboration on the development of common grade-level assessments aligned to the CCSSM, this initiative is likely to affect other important systems crucial to students' learning, including instruction, curriculum materials, teacher education, and course-taking and graduation policies.

Although not all states have adopted the CCSSM, those that have represent about 87 percent of the U.S. student population. Thus, a new day is emerging in the landscape of mathematics curriculum. Work is now underway to support teachers' learning associated with the CCSSM, to develop CCSSM-aligned curriculum materials and assessments, and to monitor the effect of common standards on the U.S. educational system.

The three chapters in this section explore the CCSSM and examine issues related to implementing common standards. Randall Charles's chapter highlights important aspects of the CCSSM, including a focus on big ideas and essential understandings.

Barbara Reys et al. offer an overview of the middle grades standards and summarize key changes in mathematics focus from current practice. Finally, Gary Martin and Eric Hart discuss current developments regarding high school mathematics standards, exploring the next steps in the evolution and use of standards to improve school mathematics.

As you read the three chapters in this section, think about the challenges and opportunities that will accompany implementing the CCSSM in your classroom, your school, and your district.

Questions for Reflection and Collective Discussion

1. If you have not already done so, review the grades K–8 sections of the *Common Core State Standards for Mathematics* (CCSSM; see http://www.corestandards.org) that are at or near the grade level you teach (e.g., if you teach grade 4, review the standards for grades 3, 4, and 5). On the basis of your review, determine what topics the CCSSM emphasizes at the grade level you teach. What, if any, changes will you need to make in your curriculum and instructional practice to align with the CCSSM? Compare your findings and ideas with those of your colleagues.

2. Charles's chapter warns teachers to "avoid interpreting every content standard only as a statement of skills students should acquire." What do you think he means by this comment? He also suggests a second pitfall to avoid: what is it?

3. Charles urges educators to "connect each content standard to the *essential understandings* and related *big ideas* students should acquire." What does this statement mean? Give an example of an essential understanding and a big idea.

4. What does "teaching mathematics through problem solving" mean? Give an example that illustrates an activity that focuses on one or more standards while also promoting problem solving.

5. Although Charles prepared his chapter for an elementary school teacher audience, what messages in the chapter apply equally to middle or high school instruction?

6. Reys et al. indicate that curricular emphasis will shift with the CCSSM for middle grades teachers and students. What are some specific examples of these changes concerning teaching fractions? Developing algebraic reasoning?

7. Reys et al. claim that the CCSSM places more emphasis on transformational geometry in the middle grades than many state standards in place prior to the adoption of the CCSSM. Is this true in your state? If so, what kinds of instructional materials and professional development will your state need to help its teachers understand and implement this new focus?

8. As you examine the standards for mathematical practice, consider how to develop these practices across grades K–12. Should particular practices be emphasized in elementary, middle, or secondary school? How can your school or district support adequate attention to developing these practices across the grades?

9. In the opening section of the Martin and Hart chapter, the authors claim that "American high school mathematics education is not meeting the needs of all our nation's students." Do you agree? Why or why not?

10. What common themes exist across all recent attempts to establish standards for middle or high school mathematics?

11. If you have not already done so, review the CCSSM standards for the algebra and geometry strands. On the basis of your review, determine what topics are emphasized most heavily in the strand you examined. What, if any, changes will you need to make in your curriculum or instructional practice to align with the CCSSM? Compare your findings and ideas with those of your colleagues.

12. At the high school level, the CCSSM includes a conceptual category for *modeling*, which the CCSSM describes as "the process of choosing and using appropriate mathematics and statistics to analyze empirical situations, to understand them better, and to improve decisions" (p. 72). Does your current high school curriculum focus on mathematical modeling? If yes, is it consistent with what the CCSSM proposes? If no, what will you need to do to focus more on CCSSM-oriented modeling in your instruction?

13. Martin and Hart emphasize the importance of aligning instructional materials with the CCSSM. When doing an alignment study, what are the advantages of an alignment mapping *instructional materials (textbooks) to standards* over the more common method of mapping *individual standards to instructional materials (textbooks)*?

14. Martin and Hart sound the following warning, applicable across grades K–12: "Although this chapter's premise, as well as the CCSSM's premise, is that a clear vision and strong standards are essential for improving mathematics education, a danger arises when standards become rigid standardization." What does this statement mean, and how might you avoid this danger?

References

Common Core State Standards Initiative. *Common Core State Standards for Mathematics*. Washington, D.C.: National Governors Association Center for Best Practices and the Council of Chief State School Officers, 2010. http://www.corestandards.org.

National Council of Teachers of Mathematics (NCTM). *Curriculum and Evaluation Standards for School Mathematics*. Reston, Va.: NCTM, 1989.

———. *Principles and Standards for School Mathematics*. Reston, Va.: NCTM, 2000.

Reys, Barbara J., ed. *The Intended Mathematics Curriculum as Represented in State-Level Curriculum Standards: Consensus or Confusion?* Charlotte, N.C.: Information Age Publishing, Inc., 2006.

Chapter 2

Making the Transition to the Common Core State Standards for Mathematics

Randall I. Charles

> What and how students are taught should reflect not only the topics that fall within a certain academic discipline, *but also the key ideas* that determine how knowledge is organized and generated within that discipline. (Common Core State Standards Initiative [CCSSI] 2010, p. 3)

FOR A NUMBER of years state mathematics content standards have been the major influence on what mathematics schools do and do not teach, emphasize, and assess, and what students and teachers believe and do not believe about mathematics. The Common Core State Standards for Mathematics (CCSSM), adopted by all but a few states, will likely continue the legacy of standards being the dominant influence on mathematics education.

Although most states have agreed to adopt the CCSSM, the next few years will be a transition period from current state standards and assessments to the CCSSM and the new aligned assessments that will emerge. This chapter will recommend two actions that schools and teachers can focus on now to make the transition to the CCSSM.

What's New?

The CCSSM comprises two sets of related standards, for mathematical content and mathematical practice. The Standards for Mathematical Content build on a rich foundation of reform in mathematics education, particularly the visions of *Principles and Standards for School Mathematics* (National Council of Teachers of Mathematics [NCTM] 2000), *Adding It Up* (National Research Council [NRC] 2001), and *Curriculum Focal Points* (NCTM 2006). At a glance, the content standards look similar to the standards found in current state content standards; they identify at each of grades K–8 the specific content to be developed at that grade.

The content standards' primary goal is to provide a roadmap to "more focused and coherent [curricula and instruction] in order to improve mathematics achievement in this country" (CCSSI 2010, p. 3). To promote focus and coherence, the content standards reflect several important shifts from most existing state standards. These shifts include (1) organizing standards into related content groups called *critical areas, domains,* and *clusters* (see fig. 2.1); (2) reducing the total number of content standards at each grade, compared to most existing state standards; (3) relocating some content to other grades; (4) eliminating some content completely; (5) adding some new content; and (6) articulating some content more specifically than state standards typically do.

Critical Area	Domain	Cluster	Standards
Critical Area 1	Domain Operations and Algebraic Thinking 4.OA	Cluster Solve Problems	Standard 1 Standard 2 Standard 3
		Cluster Factors and Multiples	Standard 4
		Cluster Patterns	Standard 5
	Domain Number Operations in Base Ten 4.NBT	Cluster Place Value	Standard 1 Standard 2 Standard 3
		Cluster Multidigit Arithmetic	Standard 4 Standard 5
Critical Area 2	Domain Number and Operations—Fractions 4.NF	Cluster Fraction Equivalence and Ordering	Standard 1 Standard 2
		Cluster Fractions from Unit Fractions	Standard 3a-d Standard 4a-c
		Cluster Decimal Notation and Fractions	Standard 5 Standard 6 Standard 7
Critical Area 3	Domain Measurement and Data 4.MD	Cluster Measurement and Conversions	Standard 1 Standard 2 Standard 3
		Cluster Represent and Interpret Data	Standard 4
	Domain Geometry 4.G	Cluster Angles and Angle Measurements	Standard 5a,b Standard 6 Standard 7
		Cluster Lines, Angles, and Shapes	Standard 1 Standard 2 Standard 3

Fig. 2.1. CCSSM content standards organization for grade 4 (Note: Critical areas' alignment with domains and clusters is not always straightforward. This figure shows one interpretation.)

The Standards for Mathematical Practice "describe varieties of expertise that mathematics educators at all levels should seek to develop in their students" (CCSSI 2010, p. 6). These abilities, processes, and dispositions enable a learner to understand mathematics and use it with understanding (see fig. 2.2). Mathematical practices can translate to students' observable verbal and written actions as they do mathematics. Because the Standards for Mathematical Practice also build on a rich foundation, including process standards (NCTM 2000) and strands of mathematical proficiency (NRC 2001), teachers and administrators will find some of the mathematical practices familiar. The CCSSM, however, expresses all the mathematical practices in more depth than previous related work; some describe crucially important elements of doing mathematics, elements that will be unfamiliar to teachers and administrators. Also, for teachers and administrators, describing mathematical practices as "standards" raises their significance. Understanding each mathematical practice's meaning and implications for teaching, learning, and assessment will likely challenge teachers more than understanding the Standards for Mathematical Content.

1.	Make sense of problems and persevere in solving them
2.	Reason abstractly and quantitatively
3.	Construct viable arguments and critique the reasoning of others
4.	Model with mathematics
5.	Use appropriate tools strategically
6.	Attend to precision
7.	Look for and make use of structure
8.	Look for and express regularity in repeated reasoning

Fig. 2.2. Mathematical practices (CCSSI 2010)

Transitioning to the Standards for Mathematical Content

Since the CCSSM Standards for Mathematical Content will influence what teachers do when they close the classroom door and teach mathematics, asking if these content standards are the right ones is essential. The litmus test would be whether the standards promote excellence in teaching and learning mathematics. Mathematics content standards communicate to students, teachers, administrators, and parents the mathematics students should know and be able to do. The CCSSM reflects this by making clear that both *conceptual understanding* and *procedural fluency* are important in teaching, learning, and assessing mathematics content. To help translate this to the classroom, some Standards for Mathematical Content explicitly call for understanding; others, for procedural fluency (see fig. 2.3).

Grade 2—2.OA.2	Fluently add and subtract within 20 using mental strategies. By the end of grade 2, know from memory all sums of two one-digit numbers.
Grade 2—2.NBT1	Understand that the three digits of a three-digit number represent the number of hundreds, tens, and ones.
Grade 4—4.NF.6	Use decimal notation for fractions with denominators 10 or 100.
Grade 4—4.NF.6	Understand a fraction a/b as a multiple of $1/b$.
Grade 6—6.NS.2	Fluently divide multidigit numbers using the standard algorithm.
Grade 6—6.SP.2	Understand that a set of data collected to answer a statistical question has a distribution that can be described by its center, spread, and overall shape.

Fig. 2.3. Sample content standards (CCSSI 2010)

We must avoid two pitfalls when transitioning from existing state content standards to the CCSSM. The first is interpreting every content standard only as a statement of skills students should acquire. The history of curriculum materials and state mathematics content standards in this country might explain teachers' inclination to interpret all CCSSM Content Standards as skill-mastery statements. For more than thirty years, U.S. textbooks have stated one or more "objectives" for each mathematics lesson. The easiest learning objectives to write are statements of skills that students should master, so most learning objectives have been of this type. Most state mathematics content standards have grown out of this tradition, so it is not surprising that most state standards are also skill-mastery statements. A finding from the Trends in International Mathematics and Science Study's video study shows that interpreting all CCSSM Content Standards as skill-mastery statements is a potential pitfall, stating, "Many teachers would like their students to understand the mathematics they study, but when asked to specify the goal for a particular lesson, most U.S. teachers ... talked about skill proficiency; few mentioned understanding" (Hiebert 2003, p. 60).

Translating standards to skill-mastery statements can have a significant negative influence on teaching and learning mathematics and on students' and teachers' beliefs about mathematics. In recent observations of classrooms and school districts, I have often noted the following significant negative influences.

- Curriculum and instruction focuses exclusively on developing skills.
- Instructors teach skills that they believe help students get correct answers without regard for whether procedures are grounded in understanding.
- Teachers and students believe that learning mathematics means only developing skill proficiency.
- Assessments evaluate only skill mastery.

- Students are uninterested in understanding.

When some content standards explicitly call for understanding and others do not, a second potential pitfall is adopting an instructional mindset of "Today I will teach for understanding; tomorrow I will teach skills." To be clear, "There is no doubt that there are rules [and skills], need for practice, and exact answers. There is a need to store facts and procedures in memory" (Hiebert et al. 1997, p. xiv). However, if one accepts learning mathematics with understanding as an essential goal, then separating acquiring skill from developing understanding is a mistake. If the CCSSM Standards for Mathematical Content are to promote excellence in teaching and learning mathematics, including higher achievement for all, they should ground every content standard in understanding. This leads to the following recommendation for transitioning to the CCSSM.

> **Transition 1:**
> Develop an understanding of the Standards for Mathematical Content. Begin by connecting each content standard to the *essential understandings* and related *big ideas* students should acquire.

Figure 2.4 shows how one grade 6 content standard can connect to several essential understandings and one big idea. Standards will improve teaching only if they are expressed clearly enough that teachers can use them for planning instruction (Hiebert and Stigler 2000). Unfortunately, many standards reduce complex mathematical ideas to succinct statements like "understand the concept of ratio." Essential understandings are topic-specific mathematical ideas and relationships students should acquire in order to learn mathematics with understanding. The essential understandings shown in figure 2.4 unpack "understand the concept of ratio" into specific ideas that need to be developed, learned, and assessed. All the essential understandings in figure 2.4 can translate to rich instructional and assessment tasks and should be part of any grade 6 instructional unit connected to standard 6.RP.1.

A *big idea* in mathematics (called a *key idea* in the CCSSM) is a statement of an idea, central to learning mathematics, that connects numerous content standards and essential understandings into a coherent whole (Charles 2005). The big idea in figure 2.4, titled Proportionality, pulls together all the essential understandings related to content standard 6.RP.1. Big ideas often cut across individual content topics and always cut across grades. For example, proportionality will connect content standards and essential understandings for many topics, such as the meanings of a fraction and multiplication, many measurement concepts and processes, and later work with linear functions, algebraic equations, and slope. Note that essential understandings and big ideas:

Common Core Standard	Some Essential Understandings	Big Idea
6.RP.1 Understand the concept of ratio and use ratio language to describe a ratio relationship between two quantities.	1. Reasoning with ratios involves attending to and coordinating two quantities. 2. Ratios are often expressed in fraction notation, although ratios and fractions do not have identical meanings. 3. Forming a ratio as a measure of a real-world attribute involves isolating that attribute from other attributes and understanding the effect of changing each quantity on the attribute of interest. 4. A number of mathematical connections link ratios and fractions. • Ratios are often expressed in fraction notation, although ratios and fractions do not have identical meanings. • Ratios are often used to make "part-part" comparisons, but fractions are not. • Ratios and fractions can be thought of as overlapping sets. • Ratios can often be meaningfully reinterpreted as fractions.	**Proportionality:** When two quantities are related proportionally, the ratio of one quantity to the other is invariant as the numerical values of both quantities change by the same factor.

Fig. 2.4. An example of connecting a content standard to essential understandings and a big idea (Lobato and Ellis 2010)

- are statements about content, not pedagogy, and do not promote a particular instructional approach;
- offer a coherent set of ideas to guide instruction and assessment; and
- derive from research when possible, the logical structure of mathematics, and teachers' experience.

Figure 2.5 is another illustration of how a big idea connects content standards and essential understandings across grades. Understanding mathematics becomes important to students when teachers make essential understandings and big ideas explicit in mathematics instruction, in student-friendly language, and when teachers encourage, value, teach, and assess understanding (Lambdin 2003). Connecting content standards and essential understandings to big ideas enables students to see mathematics as a coherent body of knowledge. You can use a number of sources to select and develop your own essential understandings and big ideas: see Barnett-Clarke et al. (2010); Blanton et al. (2011); Caldwell, Karp, and Bay-Williams (2011); Charles (2005); Clements (2004); Clements, Sarama, and DiBiase (2004); Cooney, Beckman, and Lloyd (2010); Dougherty et al. (2010); Lannin, Ellis, and Elliott (2011); Lloyd, Herbel-Eisenmann, and Star (2011); Lobato and Ellis (2010), Otto et al. (2011); Sinclair, Pimm, and Skelin (2012); and Van de Walle (2004).

Big Idea

Equivalence: Any number, measure, numerical expression, or algebraic expression (when the letter is replaced by the same number) can be represented using symbols in an infinite number of ways that have the same value.

Sample Content Standards	Some Related Essential Understandings
K.CC.3 Write numbers from 0 to 20.	Most single-digit numbers can be broken apart in more than one way, where each has the same value (e.g., 9 is 5 and 4 or 6 and 3).
	Each of the numbers 11–19 can be thought of as "10 and some more" and still have the same value (e.g., 12 is 10 and 2 more).
2.NBT.1 Understand that the three digits of a three-digit number represent amounts of hundreds, tens, and ones.	Multidigit numbers can be broken apart in more than one way, where each has the same value (e.g., 125 is 1 hundred 2 tens 5 ones or 12 tens 5 ones).
3.NBT.2 Fluently add and subtract within 1000.	Numerical expressions can be represented in more than one way, where each has the same value (e.g., $28 + 14 = 28 + 2 + 12$).
3.NF.3b Recognize and generate simple equivalent fractions.	Fractional amounts can be represented using symbols in an infinite number of ways, where each has the same value (e.g., $3/4 = 6/8 = 9/12 = \ldots$).
5.MD.1 Convert among different-sized standard measurement units in a given system.	Measurements can be represented using symbols in more than one way, where each represents the same value (e.g., 7.25 m = 725 cm).
6.EE.2 Write, read, and evaluate expressions in which letters are used for numbers.	Some algebraic expressions can be represented in more than one way, where each has the same value when the letter is replaced with the same number (e.g., $3x + 5x = [3 + 5]x$).

Fig. 2.5. An example of a big idea and related essential understandings

The goal of this first recommendation is that teachers understand the standards for mathematics content. "We understand something when we see how it is related or connected to other things we know" (Hiebert et al. 1997, p. 4). Connecting each content standard to essential understandings and big ideas is a first step in understanding these standards. The second step is analyzing the standards, essential understandings, and big ideas for the relationships and connections that exist among them. Specific possible analyses might include the following:

1. Identify relationships and connections among standards *in each grade* for every cluster, domain, and critical area.
2. Identify relationships and connections among standards *across grades* for each domain and critical area.
3. Identify relationships and connections among standards *within and across grades* for each big idea.
4. Identify the essential skills connected to particular content standards and the related essential understandings and big ideas.

Teachers can better teach for understanding when they know the essential understandings and big ideas that need development. Connecting content standards to essential understandings and big ideas is a primary step in helping teachers develop mathematics as a coherent body of knowledge and making the transition to the CCSSM.

Transitioning to the Standards for Mathematical Practice

The CCSSM presents its Standards for Mathematical Practice, like most standards, in a numbered list with descriptions accompanying each, to clarify the meaning of each mathematical practice. A pitfall to avoid when making the transition to the CCSSM is adopting an instructional plan that develops each mathematical practice in isolation from the others. Statements such as, "Today I am teaching Mathematical Practice 1; tomorrow I am teaching Mathematical Practice 2" might reflect this method.

The work students do is the most important factor influencing their understanding of mathematics. Teaching mathematics through problem solving introduces new concepts and skills to students through problem-solving tasks that contain important, embedded mathematical ideas (Lester and Charles 2003). Describing the many aspects of such teaching in this chapter is impossible. For helpful resources that do describe them, see this chapter's reference list. Students derive two learning benefits from problem-based instruction: they develop content understanding, and they gain expertise with mathematical practices.

Teaching mathematics through problem solving promotes understanding as students struggle mentally with important ideas while solving problems. "Struggle is when students expend effort to make sense of mathematics that is within reach but not yet well known; struggle is not needless frustration or challenges that are either nonsensical or overly difficult. Struggling is the opposite of being presented information to be memorized or being asked only to practice what has been demonstrated" (Bay-Williams 2010, p. 34).

Teaching mathematics through problem solving is also an effective way to develop students' proficiencies with the eight mathematical practices. The thinking and work students must do with problem-based learning calls for selecting, coordinating, implementing, and managing *multiple* mathematical practices. This leads to a second recommendation for transitioning to the CCSSM.

> **Transition 2:**
> Develop an understanding of the Standards for Mathematical Practice. Begin by making a shift to drive your daily instruction by *teaching mathematics through problem solving*, which must involve selecting, coordinating, implementing, and managing multiple mathematical practices.

Figure 2.6 shows a problem-based task used to introduce students to ratios and proportional reasoning. Predicting the kinds of thinking students might use when working on a particular problem-based task is difficult unless a teacher has used that task with many students. However, one can easily imagine how to employ multiple mathematical practices (see fig. 2.2) in working on the task in figure 2.6.

> Suppose that you have made a batch of green paint by mixing 2 cans of blue paint with 7 cans of yellow paint. What are some other combinations of numbers of cans of blue paint and yellow paint that you can mix to make the same shade of green?

Fig. 2.6. A possible problem-based learning task (Lobato and Ellis 2010)

Making the shift to teaching mathematics through problem solving is also an effective vehicle for developing teachers' understanding of the mathematical practices. Selecting a task and planning for problem-based teaching should involve analyses of the content embedded in the thinking and work students might do and of the mathematical practices students might engage with when working on the task. An important element of teaching through problem solving is the teacher facilitating students' reflection on their thinking and work. That reflection should make explicit the important mathematical content—essential understandings, big ideas, and, if appropriate, skills—students should take away from the learning experience. It should also make explicit how the students' thinking and work employed or could have employed various mathematical practices. Teachers who analyze tasks, plan for instruction, and facilitate students' reflection on content and mathematical practices will themselves develop an understanding of the Standards for Mathematical Practice.

Conclusion

"Research has shown that teachers have perhaps the strongest impact on students and their learning" (Bush and McGatha 2010, p. 19). In particular, a teacher's content knowledge is a major influence on students' learning. Connecting the CCSSM Standards for Mathematical Content to essential understandings and big ideas (transition 1) will help develop every teacher's mathematical content knowledge. Communicating to teachers not just the skills students need to master, but also the ideas and relationships students need to know and understand, will help teachers go beyond surface-level teaching of skills to developing their students' deep understandings of mathematics.

Although this chapter gave a separate recommendation for transitioning to the Standards for Mathematical Practice (transition 2), we should not separate our work on understanding mathematical practices from that on understanding content. Thinking about content, particularly essential understandings, big ideas, and mathematical practices, in isolation would be as problematic as thinking about teaching for conceptual understanding and teaching for procedural fluency as disconnected from each other. Employing mathematical practices in developing understanding and proficiency with mathematics content is essential to attaining the vision of the CCSSM.

Finally, classroom teachers and other leaders in mathematics education should not wait for the CCSSM and the related instructional materials and assessments to become operational in their districts and classrooms. They should begin the work needed to understand both the new content and mathematical practices standards immediately. In addition to the chapters in this book, numerous resources from NCTM, state departments of education, and others are becoming readily available. Professional development related to the CCSSM is beginning to flourish, too, including significant opportunities at NCTM and state mathematics conferences. Begin now making the transitions to the CCSSM that this chapter recommends. Finding a path to higher achievement for all cannot wait for policies, politics, and timelines over which we have no control.

References

Barnett-Clarke, Carne, William Fisher, Rick Marks, and Sharon Ross. *Developing Essential Understanding of Rational Numbers for Teaching Mathematics in Grades 3–5*. Essential Understanding Series. Reston, Va.: National Council of Teachers of Mathematics, 2010.

Bay-Williams, Jennifer M. "Effective Classroom Practices of Elementary School Teachers." In *Teaching and Learning Mathematics: Translating Research for Elementary School Teachers*, edited by Diana V. Lambdin and Frank K. Lester, Jr., pp. 31–35. Reston, Va.: National Council of Teachers of Mathematics, 2010.

Blanton, Maria, Linda Levi, Terry Wayne Crites, and Barbara J. Dougherty. *Developing Essential Understanding of Algebraic Thinking for Teaching Mathematics in Grades 3–5*. Essential Understanding Series. Reston, Va.: National Council of Teachers of Mathematics, 2011.

Bush, William S., and Maggie B. McGatha. "Teachers' Knowledge, Beliefs, Attitudes, and Practices." In *Teaching and Learning Mathematics: Translating Research for School Administrators*, edited by Randall I. Charles and Frank K. Lester, Jr., pp. 19–23. Reston, Va.: National Council of Teachers of Mathematics, 2010.

Caldwell, Janet H., Karen Karp, and Jennifer M. Bay-Williams. *Developing Essential Understanding of Addition and Subtraction for Teaching Mathematics in Prekindergarten–Grade 2*. Essential Understanding Series. Reston, Va.: National Council of Teachers of Mathematics, 2011.

Charles, Randall I. "Big Ideas and Understandings as the Foundation for Elementary and Middle School Mathematics." *Journal of Mathematics Education Leadership* 8, no. 1 (Spring–Summer 2005): 9–24.

Clements, Douglas H. "Major Themes and Recommendations." In *Engaging Young Children in Mathematics: Standards for Early Childhood Mathematics Education*, edited by Douglas H. Clements, Julie Sarama, and Ann-Marie DiBiase, pp. 7–72. Mahwah, N.J.: Lawrence Erlbaum Associates, 2004.

Clements, Douglas H., Julie Sarama, and Ann-Marie DiBiase, eds. *Engaging Young Children in Mathematics: Standards for Early Childhood Mathematics Education*. Mahwah, N.J.: Lawrence Erlbaum Associates, 2004.

Common Core State Standards Initiative (CCSSI). *Common Core State Standards for Mathematics*. Washington, D.C.: National Governors Association Center for Best Practices and the Council of Chief State School Officers, 2010. http://www.corestandards.org.

Cooney, Thomas J., Sybilla Beckmann, and Gwendolyn M. Lloyd. *Developing Essential Understanding of Functions for Teaching Mathematics in Grades 9–12*. Reston, Va.: National Council of Teachers of Mathematics, 2010.

Dougherty, Barbara J., Alfinio Flores, Everett Louis, and Catherine Sophian. *Developing Essential Understanding of Number and Numeration for Teaching Mathematics in Prekindergarten–Grade 2*. Essential Understanding Series. Reston, Va.: National Council of Teachers of Mathematics, 2010.

Hiebert, James. "Signposts for Teaching Mathematics through Problem Solving." In *Teaching Mathematics through Problem Solving: Prekindergarten-Grade 6*, edited by Frank K. Lester, Jr., and Randall I. Charles, pp. 53–61. Reston, Va.: NCTM, 2003.

Hiebert, James, and James W. Stigler. "A Proposal for Improving Classroom Teaching: Lessons from the TIMSS Video Study." *Elementary School Journal* 101 (September 2000): 3–20.

Hiebert, James, Thomas P. Carpenter, Elizabeth Fennema, Karen C. Fuson, Diane Wearne, Hanlie Murray, Alwyn Olivier, and Piet Human. *Making Sense: Teaching and Learning Mathematics with Understanding*. Portsmouth, N.H.: Heinemann, 1997.

Lambdin, Diana V. "Benefits of Teaching through Problem Solving." In *Teaching Mathematics through Problem Solving: Prekindergarten-Grade 6*, edited by Frank K. Lester, Jr., and Randall I. Charles, pp. 3–13. Reston, Va.: National Council of Teachers of Mathematics, 2003.

Lannin, John, Amy B. Ellis, and Rebekah Elliott. *Developing Essential Understanding of Mathematical Reasoning for Teaching Mathematics in Prekindergarten–Grade 8*. Essential Understanding Series. Reston, Va.: National Council of Teachers of Mathematics, 2011.

Lester, Frank K., Jr., and Randall I. Charles, eds. *Teaching Mathematics through Problem Solving: Prekindergarten-Grade 6*. Reston, Va.: National Council of Teachers of Mathematics, 2003.

Lloyd, Gwendolyn, Beth Herbel-Eisenmann, and Jon Star. *Developing Essential Understanding of Expressions, Equations, and Functions for Teaching Mathematics in Grades 6–8*. Essential Understanding Series. Reston, Va.: National Council of Teachers of Mathematics, 2011.

Lobato, Joanne, and Amy B. Ellis. *Developing Essential Understanding of Ratios, Proportions, and Proportional Reasoning for Teaching Mathematics in Grades 6–8*. Reston, Va.: National Council of Teachers of Mathematics, 2010.

National Council of Teachers of Mathematics (NCTM). *Principles and Standards for School Mathematics*. Reston, Va.: NCTM, 2000.

——— . *Curriculum Focal Points for Prekindergarten through Grade 8 Mathematics: A Quest for Coherence*. Reston, Va.: NCTM, 2006.

National Research Council (NRC). *Adding It Up: Helping Children Learn Mathematics*. Washington, D.C.: National Academy Press, 2001.

Otto, Albert Dean, Janet H. Caldwell, Cheryl Ann Lubinski, and Sarah Wallus Hancock. *Developing Essential Understanding of Multiplication and Division for Teaching Mathematics in Grades 3–5*. Essential Understanding Series. Reston, Va.: National Council of Teachers of Mathematics, 2011.

Sinclair, Nathalie, David Pimm, and Melanie Skelin. *Developing Essential Understanding of Geometry for Teaching Mathematics in Grades 6–8*. Essential Understanding Series. Reston, Va.: National Council of Teachers of Mathematics, 2012.

Van de Walle, John. *Elementary and Middle School Mathematics: Teaching Developmentally*. New York: Longman, 2004.

Chapter 3

Common Core State Standards for Middle Grades Mathematics:
Implications for Curriculum and Instruction

Barbara J. Reys
Barbara J. Dougherty
Travis A. Olson
Amanda Thomas

At a meeting in March 2009, a group of state governors struck an agreement. For the first time, states would collaborate on developing common learning goals (standards) for students, grades K–12, in mathematics and English language arts. The goal was to increase content standards' quality and rigor and to address the variation in students' learning opportunities inherent in a system that sets curriculum goals locally. The National Governors Association (NGA) and the Council of Chief State School Officers (CCSSO) commissioned a writing group, first to describe goals for college and career readiness and then, using those goals, to develop grades K–12 standards for mathematics and English language arts.

The authors of the mathematics standards drew on a variety of expertise and resources, including mathematical content experts, cognitive scientists, and mathematics education researchers and practitioners. They also reviewed standards from countries such as Singapore and Japan, whose students performed well on the Trends in International Mathematics and Science Study (TIMSS). Further, they studied "learning progressions detailing what is known today about how students' mathematical knowledge, skill, and understanding develop over time" (Common Core State Standards Initiative [CCSSI] 2010, p. 4). In June 2010, fifteen months after the governors' meeting, the NGA and CCSSO released the Common Core State Standards for Mathematics (CCSSM) (see http://www.corestandards.org/).

To date, all but a few states and territories—Alaska, Minnesota, Nebraska, Texas, Virginia, American Samoa, Guam, the Northern Mariana Islands, and Puerto Rico—have adopted the CCSSM. For the first time in the United States, a significant majority of schools, teachers, and students will focus on common and, in grades K–8, grade-specific learning goals for mathematics. Coupled with common grade-level assessments aligned to the CCSSM, which two state-led consortia are currently developing, this initiative will likely affect other important elements crucial to students' mathematical learning. In particular, this initiative could affect instructional techniques, curriculum materials, teacher preparation and development, and policies related to course-taking and graduation requirements.

This chapter gives an overview of the CCSSM for grades 6–8. It highlights similarities and differences between the CCSSM and both state standards, as reviewed in Reys et al. (2006), and recommendations from professional groups such as the National Council of Teachers of Mathematics (NCTM) and the American Statistical Association (ASA). It closes with a summary of work needed to ensure that we prepare and support middle grades mathematics teachers as they implement the CCSSM.

How Is the Middle Grades CCSSM Organized?

As noted earlier in this volume, the CCSSM includes two types of standards—*standards for mathematical practice* and *standards for mathematical content*. The Standards for Mathematical Practice (see table 3.1) apply to all grades, K–12, and describe different types of expertise that mathematics educators at all levels seek to develop in their students. The CCSSM based these standards on the process standards outlined in NCTM's (2000) *Principles and Standards for School Mathematics* (i.e., problem solving, reasoning and proof, communication, connections, and representation), and the strands of mathematical proficiency specified in the National Research Council's (2001) report, *Adding It Up* (i.e., adaptive reasoning, strategic competence, conceptual understanding, procedural fluency, and productive disposition).

Table 3.1
CCSSM Standards for Mathematical Practice

Number	Standards for Mathematical Practice
1	Make sense of problems and persevere in solving them
2	Reason abstractly and quantitatively
3	Construct viable arguments and critique the reasoning of others
4	Model with mathematics
5	Use appropriate tools strategically
6	Attend to precision
7	Look for and make use of structure
8	Look for and express regularity in repeated reasoning

Although the CCSSM begins with the Standards for Mathematical Practice, most of the document specifies standards for mathematical content that students in grades K–12 should study and learn. The CCSSM organizes standards for middle grades (6–8) into "domains." Table 3.2 gives general statements of the content standards for middle grades, arranged by domain.

Table 3.2
Primary Focus (Headings) of the Middle Grades Standards for Mathematical Content

Domain	Grade 6	Grade 7	Grade 8
Ratios and proportional relationships	Understand ratio concepts and use ratio reasoning to solve problems	Analyze proportional relationships and use them to solve real-world and mathematical problems	
The number system	Apply and extend previous understandings of multiplication and division to divide fractions by fractions Compute fluently with multi-digit numbers, and find common factors and multiples Apply and extend previous understandings of numbers to the system of rational numbers	Apply and extend previous understandings of operations with fractions to add, subtract, multiply, and divide rational numbers	Know that there are numbers that are not rational, and approximate them by rational numbers
Expressions and equations	Apply and extend previous understandings of arithmetic to algebraic expressions Reason about and solve one-variable equations and inequalities Represent and analyze quantitative relationships between dependent and independent variables	Use properties of operations to generate equivalent expressions Solve real-life and mathematical problems using numerical and algebraic expressions and equations	Work with radicals and integer exponents Understand the connections between proportional relationships, lines, and linear equations Analyze and solve linear equations and pairs of simultaneous linear equations
Functions			Define, evaluate, and compare functions Use functions to model relationships between quantities
Geometry	Solve real-world and mathematical problems involving area, surface area, and volume	Draw, construct, and describe geometrical figures and describe the relationships between them Solve real-life and mathematical problems involving angle measure, area, surface area, and volume	Understand congruence and similarity using physical models, transparencies, or geometry software Understand and apply the Pythagorean theorem Solve real-world and mathematical problems involving volume of cylinders, cones, and spheres
Statistics and probability	Develop understanding of statistical variability Summarize and describe distributions	Use random sampling to draw inferences about a population Draw informal comparative inferences about two populations Investigate chance processes, and develop, use, and evaluate probability models	Investigate patterns of association in bivariate data

What's Similar and What's Different from Current Practice?

In this section, we present some of the ways the CCSSM is different from, and similar to, learning goals described in state standards documents reviewed in 2006 prior to the release of the CCSSM, NCTM's (2006) *Curriculum Focal Points* (CFP), and ASA's *Guidelines for Assessment and Instruction in Statistics Education* (GAISE) (Franklin et al. 2007). We have organized this section around the content domains identified in table 3.2.

The Number System, Ratios, and Proportional Relationships

In grades 6-8, the CCSSM specifies number concepts and skills in two domains, ratios and proportional relationships (grade 6) and the number system (grades 6-8). Specific topics include the following.

Ratios and Proportional Relationships

- Use ratio concepts and ratio reasoning (grade 6)
- Analyze proportional relationships and use them to solve problems (grade 7)

The CCSSM emphasizes ratio and proportional relationships in both grades 6 and 7 with explicit connection to fractions, as previous state standards and the CFP did. In grade 6, the CCSSM calls for students to develop an understanding of ratio concepts and reasoning to describe real-world situations and solve problems, including unit rates, percents of quantities, and measurement unit conversion. The standards for grade 7 emphasize application problems that relate to intensive and extensive quantities, problem solving, and algebraic contexts. Here, the CCSSM calls for students to recognize and use proportional relationships with multiple representations that include tables, graphs on the coordinate plane, equations, diagrams, and real-world situations.

The Number System

- Divide fractions (grade 6)
- Divide multidigit numbers (grade 6)
- Compute with multidigit decimals (grade 6)
- Greatest common factor and the least common multiple (grade 6)
- Rational numbers (grades 6, 7)
  - Conceptual understanding of rational numbers (grade 6)
  - Absolute value (grade 6)
  - Compute with rational numbers (grade 7)
- Irrational numbers (grade 8)

In grade 5, prior to the middle grades, the CCSSM calls for students to learn a standard algorithm for adding, subtracting, and multiplying fractions, including mixed numbers. In particular, the CCSSM (5.NF.1) includes the following standard:

> Add and subtract fractions with unlike denominators (including mixed numbers) by replacing given fractions with equivalent fractions in such a way as to produce an equivalent sum or difference of fractions with like denominators. For example, $2/3 + 5/4 = 8/12 + 15/12 = 23/12$. (In general, $a/b + c/d = (ad + bc)/bd$.)

The CCSSM does not call for facility with division of fractions until grade 6.

The emphasis on computational fluency with fractions (addition, subtraction, and multiplication) prior to grade 6 represents a shift from the norm of previous state standards in which acquiring fluency with operations with fractions was a major focus of the middle grades (Reys et al. 2006).

As with fractions, decimal concepts and computation are introduced prior to middle grades, primarily emphasized in grades 4 and 5. In grade 5, the CCSSM specifies that students apply the four operations to decimals up to hundredths. The CCSSM extends these operations to include any multidigit decimal in grade 6 (6.NS.3):

> Fluently add, subtract, multiply, and divide multi-digit decimals using the standard algorithm for each operation.

The CCSSM calls for students in grade 7 to convert rational numbers, including integers, to decimals using long division, and in grade 8, to develop an understanding of decimal expansions and approximations for irrational numbers. For example, the CCSSM states that students should be able to do the following:

> Know that numbers that are not rational are called irrational. Understand informally that every number has a decimal expansion; for rational numbers show that the decimal expansion repeats eventually, and convert a decimal expansion which repeats eventually into a rational number. (8.NS.1)

> Use rational approximations of irrational numbers to compare the size of irrational numbers, locate them approximately on a number line diagram, and estimate the value of expressions (e.g., π^2). For example, by truncating the decimal expansion of $\sqrt{2}$, show that $\sqrt{2}$ is between 1 and 2, then between 1.4 and 1.5, and explain how to continue on to get better approximations. (8.NS.2)

Although the middle school CCSSM expectations for fractions and decimals are consistent with those of the CFP, the CCSSM introduces and calls for fluency with them earlier than many state standards. Likewise, most state standards reviewed in 2006 did not include a focus on irrational numbers in grade 8.

The CCSSM introduces the concept of integers in grade 6, focusing on the quantities the integers represent and on how one can compare their magnitudes using a number line or other representation. Then, in grade 7, computations with all rational numbers, including integers, build on these conceptual underpinnings.

Expressions and Equations

Prior to the middle grades, the CCSSM embeds algebra concepts in its operations and algebraic thinking domain. In earlier grades, algebraic reasoning focuses on generalizations of patterns and properties with numbers and on solving number sentences with missing values. In the middle grades, more specific domains relative to algebra appear. In particular, the following ideas are pertinent to algebraic thinking, reasoning, and skill development in grades 6–8.

Expressions and Equations

- Evaluate algebraic expressions (grades 6, 7)
- Fundamental properties of arithmetic (grades 6, 7)
- Linear equations and inequalities (grades 6, 7)
- Variables (grades 6, 7)
- Systems of linear equations (grade 8)
- Properties of exponents and radicals (grade 8)
- Expressions and equations, including scientific notation (grade 8)

The CCSSM standards represent a relatively formal approach to algebra, and for grades 6–8, the CCSSM is consistent with the CFP. The standards in this algebra domain, however, represent a shift from current practice as specified in many state standards. For example, the following CCSSM standard (4.OA.3) states how students should be able to use variables by grade 4:

> Solve multistep word problems posed with whole numbers and having whole-number answers using the four operations, including problems in which remainders must be interpreted. Represent these problems using equations with a letter standing for the unknown quantity. Assess the reasonableness of answers using mental computation and estimation strategies including rounding.

Thus, these earlier experiences may provide the necessary background knowledge for grade 6 students to regularly use variables to represent unknowns in expressions and equations.

Both the CCSSM and the CFP call for students to use the fundamental properties of arithmetic to manipulate expressions and equations and to create and justify equivalent forms of expressions. The two sets of standards develop these properties, considered central to computational fluency, in the early grades through a generalized arithmetic approach.

Modeling real-world and mathematical situations with expressions and equations requires students to apply the concept of variable and ideas of equivalency, since an expression's or equation's form can more explicitly show relationships among quantities or variables. The CFP does not specifically treat modeling until grade 8, whereas the CCSSM has explicit standards for modeling with symbolic statements in grade 7, such as the following (7.EE.4):

> Use variables to represent quantities in a real-world or mathematical problem, and construct simple equations and inequalities to solve problems by reasoning about the quantities.

The CFP and the CCSSM both include the expectation that, by the end of grade 8, students solve systems of linear equations through multiple methods that might include graphing, tables, or algebraic techniques. Both also call for students to determine if a system of two linear equations has only one solution, infinitely many solutions, or no solution by generalizing characteristics of equations and their associated graphs. In 2006, however, only ten state standards identified solving systems of equations as a goal in grade 8 or earlier (Newton, Larnell, and Lappan 2006). Including this goal in the CCSSM is likely to represent a new emphasis in many middle school mathematics programs.

One difference between the CFP's and CCSM's recommendations relates to inequalities. The CCSSM develops inequalities concepts and skills at the same time as equations. For example, the CCSSM calls for students in grade 6 to model and solve mathematical and real-world situations that can be represented by equations and inequalities. Most state standards reviewed in 2006 introduced inequalities later than equations, most often in grades 7 and 8 (Newton, Larnell, and Lappan 2006).

Functions

One of three crucial focus areas for grade 8 is function, that is, understanding the concept of a function and using functions to describe quantitative relationships. Broad goals include the following:

- Define, evaluate, and compare functions (grade 8)
- Use functions to model relationships between quantities (grade 8)

Both the CFP and the CCSSM include representing functions in multiple ways, including symbolic, natural language, tables, and graphs, at grade 8. Although neither calls for students to use specific function notation, understanding what a function is and what relationships of quantities functions represent is essential. Both documents require that students identify the rate of change consistent between a function's two variables as the slope of a line.

The CCSSM standards for grade 8, however, emphasize a deep understanding of functions. For example, in addition to understanding that that the slope of a line is the relationship between the change in the values of one variable (x-coordinate) to the change in the other variable (y-coordinate), the CCSSM expects students to explain why the slope of a line is the same, regardless of the points selected on the line. The specific focus in the CCSSM links to using similar triangles and results in deriving the slope-intercept form of an equation.

A more subtle difference between the CFP's and CCSSM's expectations relates to generalizing characteristics of functions that one can apply in different ways. For example, the CCSSM calls for students to sketch graphs that represent functions described more qualitatively, such as with natural language in the following (8.F.5):

> Describe qualitatively the functional relationship between two quantities by analyzing a graph (e.g., where the function is increasing or decreasing, linear or nonlinear). Sketch a graph that exhibits the qualitative features of a function that has been described verbally.

Students will need opportunities to express generalizations that describe the characteristics of graphs if they know the rule. Conversely, students need similar opportunities to describe the characteristics of a symbolic rule given a graph.

Geometry

Beginning with grade 6, the CCSSM includes measurement standards in the geometry domain rather than as a separate domain. Important topics in this domain include the following:

- Area, volume, surface area (grades 6–8)
- Coordinates on a plane (grade 6)
- Scale drawings and constructions (grade 7)
- Angles (grades 7–8)
- Transformations (rigid and nonrigid motions) (grade 8)
- Pythagorean theorem (grade 8)

Area, Volume, and Surface Area

The CCSSM treatment of area and volume is similar to what most state standards and the CFP include. That is, grade 6 emphasizes the areas of triangles and special quadrilaterals. The standards include statements to "compose and decompose" shapes into other known shapes and to "pack" right rectangular prisms with unit cubes, emphasizing the need to build new knowledge from existing understanding. Likewise, grade 6 specifies finding the surface area of three-dimensional figures by representing figures with "nets."

At grade 7, the CCSSM calls for students to solve problems involving area, volume, and surface area and to "know the formulas for the area and circumference of a circle" (p. 50). At grade 8, students are expected to solve problems involving volumes of cylinders, cones, and spheres. In every instance, grades 6–8, standards for measuring circles and polygons include attention to applying techniques to solve both real-world mathematical problems and problems without a context.

One example of the CCSSM's specificity regarding particular models or derivations occurs in grade 7, where students are to "give an informal derivation of the relationship between the circumference and area of a circle" (p. 50). For example, students might relate a circle's area to its circumference, as shown in figure 3.1 below, by dissecting a circle with circumference C to form an approximation to a parallelogram.

Coordinates on a Plane

The CCSSM calls for grade 5 students to graph points in the first quadrant on a coordinate plane. In grade 6, the expectations broaden to include graphing points in all four quadrants. Moreover, the CCSSM specifically refers to using the coordinate plane to plot points that represent vertices of polygons. Using these coordinates, students determine the side lengths of polygons. These grade 6 CCSSM expectations are consistent with most state standards reviewed in 2006.

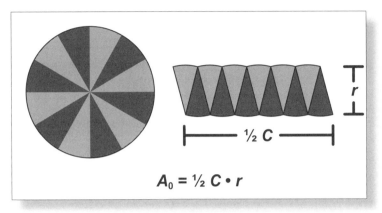

Fig. 3.1. Relating a circle's area to its circumference

Scale Drawings and Constructions

Like most state standards and the CFP, the CCSSM includes the expectation at grade 7 for students to solve problems involving scale drawings (e.g., computing lengths and areas from scale drawings and reproducing a scale drawing at a different scale). Less common in the state standards at grade 7, and absent from the CFP, is the expectation to "draw geometric shapes with given conditions" (p. 50) as a means to develop generalizations about various shapes. Another example of a new item in the CCSSM, absent from state standards or the CFP, is the expectation to "describe the two-dimensional figures that result from slicing three-dimensional figures" (p. 50).

Angles

The CCSSM at grade 7 calls for students to learn the terms and concepts associated with supplementary, complementary, vertical, and adjacent angles, and to use those concepts to determine the measures of unknown angles in figures. As with the CFP, in the CCSSM's grade 8, students extend their knowledge to angle sums of polygons, exterior angles of triangles, angles created by parallel lines cut by a transversal, and relationships among angles in similar triangles.

Transformations

The most noticeable difference between state standards reviewed in 2006 and the CCSSM at the middle grades is the latter's emphasis on transformational geometry in grade 8. The following four CCSSM standards specify these learning goals:

1. Verify experimentally the properties of rotations, reflections, and translations:
    a. Lines are taken to lines, and line segments to line segments of the same length.
    b. Angles are taken to angles of the same measure.
    c. Parallel lines are taken to parallel lines. (8.G.1)

2. Understand that a two-dimensional figure is congruent to another if the second can be obtained from the first by a sequence of rotations, reflections, and translations; given two congruent figures, describe a sequence that exhibits the congruence between them. (8.G.2)
3. Describe the effect of dilations, translations, rotations, and reflections on two-dimensional figures using coordinates. (8.G.3)
4. Understand that a two-dimensional figure is similar to another if the second can be obtained from the first by a sequence of rotations, reflections, translations, and dilations; given two similar two-dimensional figures, describe a sequence that exhibits the similarity between them. (8.G.4)

Most state standards placed similar learning goals among high school (geometry) standards. Moreover, state standards reviewed in 2006 introduced transformational geometry (rigid motions) informally, and often superficially, in the early grades.

Pythagorean Theorem

The CCSSM's development of the Pythagorean theorem differs from that of previous state standards, particularly relating to its emphasis on validating the theorem. For example, the CCSSM specifies that students should be able to "Explain a proof of the Pythagorean Theorem and its converse." (8.G.6)

Only thirteen state standards examined in 2006 called for students to verify, develop, validate, justify, derive, or prove the Pythagorean theorem in grade 8. No states specified formal or informal proof of the Pythagorean theorem prior to grade 8. In fact, the state standards most commonly placed this expectation among high school learning goals, in the geometry domain. Thus, the CCSSM expectation to explain a proof of the Pythagorean theorem and its converse in grade 8 represents a shift to earlier grades.

Statistics and Probability

Unlike most state standards reviewed in 2006, the CCSSM includes very little attention in grades K–5 to statistics and probability. In fact, the CCSSM has no statistics and probability domain in grades K–5. Instead, it includes expectations involving data representation and analysis in the measurement and data domain.

The CCSSM begins a major focus on statistics and probability in the middle grades and extends it into high school. It organizes standards in this domain at grades 6–8 as follows:

- Variability and spread (grades 6, 8)
- Measures of center (grades 6–7)
- Displaying numerical data in plots (grades 6, 8)
- Examining a sample of the population/random sampling (grade 7)
- Comparing two populations (grade 7)
- Probability of an event is between 0 and 1, inclusive (grade 7)
- Law of large numbers (grade 7)

- Develop probability models and compare probabilities found using models (grade 7)
- Probabilities of compound events (grade 7)
- Sample space (grade 7)
- Patterns in bivariate data (grade 8)
- Use equations of linear models to solve contextualized problems (grade 8)

Since ASA's GAISE report has influenced the content and sequence of many state standards, we examined the extent to which the CCSSM standards for middle grades aligned with the GAISE Level B recommendations. Overall, the CCSSM draws from, and generally aligns with, Level B GAISE recommendations, particularly concerning the focus on collecting and analyzing data and interpreting results. For example, the CCSSM's grade 6 standard that states that students "Display numerical data in plots on a number line, including dot plots, histograms, and box plots" (p. 45) relates directly to a Level B description in the GAISE document that calls for students to "use more sophisticated tools for summarizing and comparing distributions, including: Histograms, … , and box plots" (p. 38). However, GAISE places more emphasis than the CCSSM on students being prepared to formulate questions about their statistical examinations.

The CCSSM standards in this domain are also generally consistent with the CFP's recommendations. For example, the CCSSM's grade 8 statistics and probability standards focus on interpreting bivariate data. Likewise, the CFP recommends that "students make scatterplots to display bivariate data, and they informally estimate lines of best fit to make and test conjectures" (p. 40). This language is nearly identical, albeit considerably more condensed, than the CCSSM's four grade 8 statistics and probability standards.

Differences exist between the CCSSM's content and that of state standards reviewed in 2006. For example, the CCSSM includes attention to a variety of data displays (histograms, box plots) at grade 6, whereas most states included this focus in grades 7 and 8. The state documents also specified a wider range of representations, such as circle graphs, stem-and-leaf plots, histograms, box plots, and scatterplots (Newton, Hovarth, and Dietiker 2011); the CCSSM does not specify attention to circle graphs or stem-and-leaf plots. Likewise, most state standards reviewed in 2006 focused explicitly on the fundamental counting principle as a method for finding a complete sample space. This concept's grade placement in state standards ranged from grade 6 to 8. The CCSSM, however, does not mention this concept explicitly in its middle grades or high school standards.

Dingman and Tarr (2011) identify other important probability concepts that the CCSSM, the CFP, and state standards reviewed in 2006 addressed differently. For example, the state standards vary in their placement of the sample space concept, ranging from grades 5 to 7, whereas the CCSSM places it in grade 7. Likewise, state standards placed the concept that the probability of an event lies inclusively between 0 and 1 somewhere in grades 4–8, depending on the state, whereas the CCSSM has it in grade 7.

What's Needed? Implications and Challenges in Implementing the CCSSM

In some domains, the CCSSM represents a significant shift from current practice in middle school mathematics teaching and learning. For example, little attention is placed on computation of whole numbers, fractions, or decimals in the middle grades. Instead, more emphasis is placed on geometry, algebra, and statistics and probability. Specifically, throughout the middle grades, the CCSSM includes increased attention to the following:

- algebra, including significant attention to algebraic skills (e.g., equation solving) and introducing functions in grade 8;
- transformational geometry;
- verification through formal processes (e.g., demonstrating a proof of the Pythagorean theorem); and
- statistics and probability.

Curriculum materials development, instructional planning, and professional development will need to be reorganized to allow for the differences between the CCSSM and the state standards that preceded them.

The CCSSM represents an acceleration of some important ideas across grades K–12. That is, the CCSSM calls for students to learn certain concepts and skills earlier than specified in prior state standards documents. Schools will need to have plans in place to transition students and teachers to the CCSSM. In the first years of implementation, this may result in identifying more students for support in programs, such as Response to Intervention, to help shore up areas where students may not have had opportunities to learn the needed mathematics, and to help them access the new content.

Students identified as struggling learners, with an Individualized Education Program constructed with splinter skills, have an additional challenge. The CCSSM standards apply to all students; thus, any changes in assessments and requirements for graduation or course taking will affect students whose instruction may have insulated them from sophisticated mathematics. The CCSSM standards will necessitate special education and general education teachers working together to create a robust program to support students' learning.

Parents should expect that their children will experience mathematics, and be expected to demonstrate understandings of it, differently from how the parents themselves may have experienced it. Although the CCSSM has received much press, parents will still need opportunities for help in understanding what the CCSSM expects of their children, and how that translates into expectations of effort, homework assignments, classroom instruction, and classroom, district, and state assessments.

The need for curriculum materials that build understanding over time, rather than present mathematics as a series of disconnected, one-shot lessons, is not particular to the middle grades. Since both the CCSSM's content and mathematical practices are new, at least at some grades, existing curriculum materials need major revisions. Schools will also need good tools for analyzing curriculum materials in order to make sound judgments about their alignment to the CCSSM.

Although not all states have adopted the CCSSM, those that have represent nearly 90 percent of the U.S. student population. Thus, a new day is emerging in the landscape of mathematics education. Work is now underway to introduce and support CCSSM-associated learning for teachers, to develop CCSSM-aligned curriculum materials and assessments, and to study common core standards and their effect on the U.S. educational system.

References

Common Core State Standards Initiative (CCSSI). *Common Core State Standards for Mathematics.* Washington, D.C.: National Governors Association Center for Best Practices and the Council of Chief State School Officers, 2010. http://www.corestandards.org.

Dingman, Shannon, and James E. Tarr. "An Analysis of K-8 Probability Standards." In *Variability is the Rule: A Companion Analysis of K–8 Mathematics Standards,* edited by John P. Smith III, pp. 161–92. Charlotte, N.C.: Information Age Publishing, 2011.

Franklin, Christine, Gary Kader, Denise Mewborn, Jerry Moreno, Roxy Peck, Mike Perry, and Richard Scheaffer. *Guidelines for Assessment and Instruction in Statistics Education (GAISE) Report: A Pre-K-12 Curriculum Framework.* Alexandria, Va.: American Statistical Association, 2007.

National Council of Teachers of Mathematics (NCTM). *Principles and Standards for School Mathematics.* Reston, Va.: NCTM, 2000.

———. *Curriculum Focal Points for Prekindergarten through Grade 8 Mathematics: A Quest for Coherence.* Reston, Va.: NCTM, 2006.

National Research Council. *Adding It Up: Helping Children Learn Mathematics.* Washington, D.C.: National Academy Press, 2001.

Newton, Jill, Aladar K. Hovarth, and Leslie Dietiker. "The Statistical Process: A View across K-8 State Standards." In *Variability Is the Rule: A Companion Analysis of K–8 Mathematics Standards,* edited by John P. Smith III, pp. 119-60. Charlotte, N.C.: Information Age Publishing, 2011.

Newton, Jill, Gregory V. Larnell, and Glenda Lappan. "Analysis of K-8 Algebra Grade-Level Learning Expectations." In *The Intended Mathematics Curriculum as Represented in State-Level Curriculum Standards: Consensus or Confusion?* edited by Barbara J. Reys, pp. 59–88. Charlotte, N.C.: Information Age Publishing, 2006.

Reys, Barbara J., Shannon Dingman, Travis Olson, Angela Sutter, Dawn Teuscher, and Kathryn Chval. "Analysis of K-8 Number and Operation Grade-Level Learning Expectations." In *The Intended Mathematics Curriculum as Represented in State-Level Curriculum Standards: Consensus or Confusion?,* edited by Barbara J. Reys, pp. 15-58. Charlotte, N.C.: Information Age Publishing, 2006.

Chapter 4

Standards for High School Mathematics in the Common Core State Standards Era

W. Gary Martin
Eric W. Hart

> Compare your own experience of learning algebra with Bertrand Russell's recollection:
>
>> I was made to learn by heart: "The square of the sum of two numbers is equal to the sum of their squares increased by twice their product." I had not the vaguest idea what this meant and when I could not remember the words, my tutor threw the book at my head, which did not stimulate my intellect in any way.
>
> Are things really any different today? (Lockhart 2002, p. 16)

Over the past several years, a clarion call has sounded for improvement in U.S. high school mathematics classrooms. Report after report describe the problem and propose solutions, often involving standards of some sort. A rather unexpected flurry of activity by the National Governor's Association and the Council of Chief State School Officers has now operationalized this energy for change in the Common Core State Standards (CCSS) (Common Core State Standards Initiative [CCSSI] 2010a), which includes standards for high school mathematics. In this chapter, we give a brief overview of the current state of standards for high school mathematics—why we need them, how the Common Core State Standards for Mathematics (CCSSM) builds on past recommendations for change, and what opportunities and pitfalls confront us as we move forward to improve high school mathematics education in the CCSSM era.

Why Standards?

Years of data have shown that American high school students are not performing up to expectations in mathematics. Only a quarter of U.S. twelfth graders scored "proficient" or above on the 2009 National Assessment of Educational Progress's (NAEP) main mathematics assessment, and significantly smaller percents of African American and Hispanic students achieved proficiency than white students (National Center for Educational Statistics [NCES] 2010). Although this is a small but significant improvement from the previous assessment in 2005, the number of students performing at the "advanced" level did not change. Moreover, NAEP's long-term trend assessment shows that 17-year-old students' scores virtually have not changed since 1990, despite significant improvements for students ages 9 and 13 (NCES 2009). A 2004 study of high school seniors found that only 35 percent "showed an understanding of intermediate-level mathematical concepts," and only 4 percent "exhibited a mastery of complex multistep word problems" (NCES 2005, p. 5).

International tests show similar disappointing results. The Programme for International Student Assessment (PISA) assesses "an individual's capacity to identify and understand the role that mathematics plays in the world, to make well-founded judgments and to use and engage with mathematics in ways that meet the needs of that individual's life as a constructive, concerned and reflective citizen" (PISA 2010, p. 32). U.S. 15-year-olds scored significantly lower than the average for industrialized nations, with the United States ranking twenty-fifth out of the thirty-three participating industrialized countries (PISA 2010). About three-quarters of U.S. students scored a level 3 or below, out of 6, indicating that they can, at best, work with "clearly described procedures" and apply "simple problem solving strategies," but cannot deal with more complex problem-solving situations.

Many other indicators and reports show that U.S. high school graduates are not adequately prepared in mathematics. For example, the National Commission on the High School Senior Year (NCHSSY) found that 24 percent of all college students require remediation in mathematics (NCHSSY 2001). A more recent study (Aldeman 2010) showed that Florida's colleges placed 31 percent of their students in remedial courses. Untold additional students require remediation in the workplace. Murray (2009) put the price tag on workplace remediation at $107–$447 million in California alone. A report from the American Diploma Project (ADP 2004, p. 1) dismally asserts that

> for too many graduates, the American high school diploma signifies only a broken promise. While students and their parents may still believe that the diploma reflects adequate preparation for the intellectual demands of adult life, in reality it falls short of this common sense goal.

Widespread concern about staying globally competitive also exists. For example, a report from the National Academy of Sciences voices concern that "the scientific and technological building blocks critical to our economic leadership are eroding at a time when many other nations are gathering strength" (Committee on Prospering in the Global Economy of the 21st Century 2007) and characterized the situation in mathematics and science education as a "gathering storm." A 2010 follow-up report by the same committee characterized this storm as "rapidly approaching category 5," the most severe level possible (Members of the 2005 "Rising Above the Gathering Storm" Committee 2010).

The inevitable conclusion from all these assessments and reports is that American high school mathematics education is not meeting the needs of all our nation's students. To achieve our educational goals, we need a clear vision, with specified objectives and expectations.

Past and Current Standards Compared

Many groups have been engaged in trying to set standards, provide guidelines, and make recommendations for high school mathematics. In this section, we examine past recommendations in order to identify general themes, similarities, and differences, and to see how the CCSSM builds on these recommendations.

Problem Solving: A Consistent Theme and Goal

One of the most prominent general recommendations is for "problem solving." The National Council of Teachers of Mathematics (NCTM) has consistently called for a focus on problem solving, beginning with *An Agenda for Action*, in which the first of eight recommendations is that "problem solving be the focus of school mathematics" (NCTM 1980). NCTM's first *Standards* document (NCTM 1989) elaborated this theme, with its Standard 1 for high school mathematics, "Mathematics as Problem Solving" and the statement that "Mathematical problem solving, in its broadest sense, is nearly synonymous with doing mathematics" (NCTM 1989, p. 137). In NCTM's *Principles and Standards for School Mathematics* (2000, p. 29), problem solving is the first "process standard." This consistent focus on problem solving represents an attempt by NCTM to shift high school mathematics away from teaching mathematical procedures exclusively, or even procedures and concepts followed by applications and problem solving, to teaching mathematics, both procedures and concepts, for and *through* problem solving, a perspective that Schoen (2003) explicates.

In 2009, NCTM made recommendations specifically for high school mathematics in *Focus in High School Mathematics: Reasoning and Sense Making* (FHSM). This volume's emphasis on mathematical reasoning builds on NCTM's continuing focus on problem solving and the process standards, arguing that reasoning and sense making are the foundations for the other processes and that those processes are all "manifestations of making sense of mathematics and of reasoning" (NCTM 2009, p. 5). The document goes on to describe "reasoning habits" that illustrate "the types of thinking that should become routine and fully expected in the classroom culture of all mathematics classes across all levels of high school" (p. 9).

Attention to problem solving is also called for at the college level. The Mathematical Association of America (MAA) has issued several recent reports (MAA 2004, 2007a, 2007b) recommending a focus on problem solving and mathematical modeling in undergraduate mathematics courses. From the perspective of the world of work, the Partnership for 21st Century Skills (2007) conducted surveys of voters and employers in 2007 and 2006, respectively, and found that "voter attitudes have shifted away from the 'back to basics' movement that was a strong theme during the 1990s" (p. 3), and "a virtually unanimous 99 percent of voters say that teaching students a wide range of 21st century skills—including critical thinking and problem-solving skills … —is important to our country's future economic success" (p. 1). Also, "while the 'three Rs' are still fundamental to any new workforce entrant's ability to do the job, employers emphasize that applied skills like Teamwork/Collaboration and Critical Thinking/Problem Solving are 'very important' to success at work" (Partnership for 21st Century Skills 2006, p. 9).

Countries that outperform the United States on high school mathematics assessments also emphasize the importance of problem solving. For example, the Singapore Ministry of Education (2006) states, "Mathematical problem solving is central to mathematics learning" (p. 2), as shown in the Ministry's current mathematics framework (fig. 4.1).

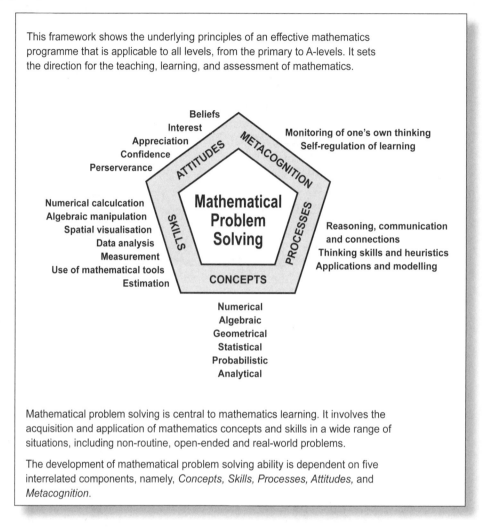

Fig. 4.1. Singapore school mathematics framework

The CCSSM builds on these themes of problem solving, reasoning, and sense making in its Standards for Mathematical Practice (see fig. 4.2). Although the CCSSM standards use different language—for example, more mathematical language such as "structure" rather than the language of "patterns" prevalent in past recommendations—and place more emphasis on abstraction and precision, these standards nevertheless explicitly build on NCTM's process standards (cf. CCSSI 2010a, p. 6). A report released by NCTM (2011) demonstrates the connections among the CCSSM mathematical practices, the *Principles and Standards* process standards (NCTM 2000), and the reasoning habits from FHSM (NCTM 2009). It is particularly important to note that CCSSM insists that the Standards for Mathematical Practice, along with all the content standards, must be included in student assessments.

1.	Make sense of problems and persevere in solving them
2.	Reason abstractly and quantitatively
3.	Construct viable arguments and critique the reasoning of others
4.	Model with mathematics
5.	Use appropriate tools strategically
6.	Attend to precision
7.	Look for and make use of structure
8.	Look for and express regularity in repeated reasoning

Fig. 4.2. CCSSM Standards for Mathematical Practice (CCSSI 2010)

The Breadth of the Curriculum

Standards documents offer similar yet different recommendations regarding the curriculum's breadth. Prior to the CCSSM's release, a number of national recommendations for essential high school mathematics content appeared in major reports from NCTM (2000, 2009); Achieve, Inc. (2007a, 2007b, 2007c), a not-for-profit group formed by governors and business leaders to "raise academic standards and achievement"; College Board (2006, 2007), a not-for-profit association "whose mission is to connect students to college success and opportunity"; and the American Statistical Association (ASA) (Franklin et al. 2007).

With the exception of the ASA recommendations, which focus solely on statistics and probability, all the reports have the same general content, with minor variations. All include strands in algebra, geometry, and statistics and probability. NCTM (2000) has additional strands in measurement and in number and operations, which reflect its standards' grades K–12 range. The other documents subsume these strands elsewhere. FHSM (NCTM 2009), which specifically focuses on high school mathematics, refines algebra into two separate strands—symbolic algebra and functions—and combines number and measurement to form a strand focusing on quantity. Achieve recommends an additional strand in discrete mathematics; the other documents include some discrete mathematics topics, such as vertex-edge graphs, recursion, and counting, in other strands.

Note that all these recommendations call for a rich mathematical sciences curriculum that includes substantial attention to content beyond just algebra and geometry. This broad focus is consistent with the recommendation from Steen (2006, p. 93) that the high school mathematics curriculum should consider equally the needs of all disciplines and careers in which mathematical tools are used, as well as the quantitative aspects of general education. Preparing all students to squeeze through the calculus filter is neither appropriate nor effective as a way to meet the mathematical expectations of higher education.

The recommendations of the CCSSM reflect this consensus on the need for a broad mathematical preparation for students. The CCSSM includes six "conceptual categories" for high school mathematics, which look quite similar to the strands in the other recommendations. Following the FHSM's lead, the CCSSM separates algebra into two conceptual categories, one on symbolic algebra and the other on functions. The CCSSM also combines "number and quantity" into a single conceptual category. It includes some discrete mathematics topics, such as recursion and counting, within its other conceptual categories while neglecting others, such as vertex-edge graphs. However, the CCSSM takes one significant step beyond the other recommendations: it includes "modeling"—described as "the process of choosing and using appropriate mathematics and statistics to analyze empirical situations, to understand them better, and to improve decisions" (CCSSI 2010, p. 72)—as an additional conceptual category. Although the CCSSM suggests no specific content standards for modeling, the document uses a "star" to denote standards in the other conceptual categories that relate to modeling. This attention on modeling throughout the other content recommendations emphasizes its importance in the high school curriculum.

The Organization of Recommendations

Despite the general similarities, significant differences also exist among the various sets of recommendations, in organization, specificity, and types of recommendations. NCTM (2000, 2009) and ASA (Franklin et al. 2007) organize their recommendations by grade span, independent of specific courses, whereas the Achieve and the College Board documents recommend teaching specific content in each high school course, either in the usual single-subject courses (algebra 1, geometry, algebra 2) or in courses in an integrated sequence that teach the mathematical strands together in a connected, coherent manner.

Like the NCTM and ASA documents and unlike Achieve and College Board, the CCSSM lists standards to be covered by the end of high school instead of giving explicit, course-by-course descriptions of how one should cover the material. A follow-up document to the CCSSM, *Appendix A: Designing High School Mathematics Courses Based on the Common Core State Standards* (CCSSI 2010b), gives optional recommendations for the content and organization of high school courses. This appendix describes two types of organization: single-subject, typical in the United States; and integrated, which is most common internationally. It gives "model pathways" for both "compacted" (i.e., accelerated but with no content omitted) and noncompacted sequences of courses in each type of organization. *Appendix A* stresses that "curriculum designers may create alternative model pathways with altogether different organizations" and that its four model course pathways are "intended to contribute to the conversations around assessment and curriculum design, rather than end them" (p. 4).

The Specificity of Recommendations

The recommendations from different groups also differ in the level of detail. For example, *Principles and Standards* (NCTM 2000) sets forth fifteen general "expectations" for algebra across the high school years, whereas Achieve (2007b) recommends twenty-four objectives for Algebra 1 alone, many of them with several subobjectives. The College Board standards fall somewhere in between, listing course-level recommendations with far fewer objectives than Achieve. FHSM (NCTM 2009) does not include any specific content recommendations; instead, its goal is to demonstrate how reasoning and sense making can become the central focus for high school mathematics. Accordingly, it delineates three to five "key elements" in each strand in which students might develop reasoning and sense making productively.

Like the Achieve and College Board standards, the CCSSM puts forward detailed descriptions of what students should know and be able to do. For example, the two conceptual categories (i.e., strands) of algebra and functions involve fifty-five standards spread across all years of the high school curriculum. However, the CCSSM requires only about forty-five of them for all students to ensure that they are "college and career ready." The document specifies the remainder as additional content, denoted with a "+," that students "should learn in order to take advanced courses such as calculus, advanced statistics, or discrete mathematics" (p. 57). Overall, the CCSSM presents six conceptual categories that one may generally think of as strands, such as "Algebra." Each conceptual category contains four to six domains, such as "reasoning with equations and inequalities." Each domain organizes its standards into clusters, such as the five standards specified in the cluster "solve systems of equations," two of which the cluster labels "+," for additional content.

Types of Recommendations

Concerning the types of recommendations made, quite different views of the content emerge from the different sets of recommendations. To illustrate this point, we will examine some examples from algebra and functions. One of the algebra standards from *Principles and Standards* (NCTM 2000, p. 296) recommends that students should "understand relations and functions and select, convert flexibly among, and use various representations for them." The description of FHSM's first key element in its *reasoning with functions* strand, "using multiple representations of functions," reflects this perspective (NCTM 2009, p. 41):

> Representing functions in various ways, including tabular, graphic, symbolic (explicit and recursive), visual, and verbal; making decisions about which representations are most helpful in problem-solving circumstances; and moving flexibly among those representations.

In contrast, although the Achieve (2007b) objectives address various representations of functions in many places, such as "identify whether a table, graph, formula, or context suggests a linear, direct proportional, or reciprocal relationship" (p. 6), none explicitly addresses coordinating and converting among the different representations.

Hirsch et al. (2007) analyzed some of the differences in prepublication drafts of the Achieve and College Board standards. They concluded that, for the treatment of algebra, "Achieve focuses on the structure of algebra with an emphasis on developing students' procedural fluency and their ability to characterize and understand quantitative relationships that can be modeled by functions," whereas "College Board characterizes algebra as a way of using symbols to represent mathematical and real-world situations and functions as a way to model patterns of change. The focus is on symbol sense and symbolic reasoning" (p. 48). An analysis of verbs used to state the recommendations indicates that the College Board standards may express more high-level cognitive demands (Hirsch et al. 2007). Despite different emphases, both documents clearly promote students' learning of a rich curriculum that encourages deep understanding of important mathematics, a perspective consistent with the NCTM and ASA recommendations as well.

Once again, the CCSSM generally falls in line with other proposed standards, in this instance also addressing the need for students to acquire a deep understanding of high school mathematics. To follow our example of "representations of functions" begun above, the CCSSM has a cluster of three standards, with seven substandards, titled "Analyze functions using different representations" in the Interpreting Functions domain of the Functions conceptual category. The stems of those three standards follow:

- "Graph functions expressed symbolically and show key features of the graph, by hand in simple cases and using technology for more complicated cases." (Five substandards describe particular classes of functions to which this standard applies.)
- "Write a function defined by an expression in different but equivalent forms to reveal and explain different properties of the function." (Two substandards describe particular types of algebraic manipulations that students should use.)
- "Compare properties of two functions, each represented in a different way (algebraically, graphically, numerically in tables, or by verbal descriptions)." (An example is provided.)

Although the CCSSM standards are quite specific, the verbs used suggest that the CCSSM intends a deeper treatment of mathematics beyond a simple focus on skills. Moreover, the structure of domains and clusters may help teachers make higher-order connections among the standards, rather than teach them as discrete topics. This is certainly the intent of the CCSSM; as stated in its introduction, where it explains the structure of domains and clusters, "mathematics is a connected subject."

The CCSSM's Prospects and Pitfalls

We have made the case that a pressing need exists to improve high school mathematics education in the United States, and that a clear vision with strong standards can help us meet this need. Many recent attempts have tried to provide the right vision and standards. We have attempted to show through our analysis that the CCSSM, the latest attempt, naturally fits in with and builds on previous recommendations. It explicitly supports the importance of mathematical problem solving, reasoning, and process through its Standards for Mathematical Practice. It supports a broad view of the high school mathematics curriculum, beyond the typical focus on algebra and geometry, and includes substantive attention to areas such as statistics and probability for all students. Although it puts forth quite detailed content standards, those standards arguably call for a deeper understanding of mathematics, and the structure of the standards may obviate against a detrimental "checklist" mentality that may naturally follow such detailed standards.

Perhaps the biggest contrast between the CCSSM and preceding sets of recommendations is that the CCSSM will have more force of accountability. At the time of writing, 90 percent of the U.S. states and 33 percent of the territories have adopted the CCSSM as their official state standards. Although minor variations are permissible, consisting of up to 15 percent additional content, states that decide to adopt the CCSSM standards must do so verbatim. State standards are the basis, at least in principle, for state accountability systems under the No Child Left Behind Act (NCLB), and states are increasingly moving to include end-of-course assessments as a part of NCLB compliance (Great Schools 2010). We can, therefore, expect states to take the CCSSM standards much more seriously than those of any of the previous documents, which were nonbinding recommendations for what should be happening across the nation. Furthermore, major assessment initiatives are in progress whereby coalitions of states are developing standardized high-stakes tests based on the CCSSM, including for high school (U.S. Department of Education 2010).

Thus, the CCSSM has the power to bring about substantial change in American high school mathematics education, change unprecedented in its nationwide consistency and leveraged accountability. In short, the *Common Core State Standards for Mathematics* document, despite the phrase "State Standards," surprisingly could provide the first-ever standards to be nationwide in their impact across the United States. Unfortunately, a number of significant pitfalls stand in the way of making significant progress based on the CCSSM. Some of these pitfalls are more immediate, reflecting the difficulty of implementing a new vision for high school mathematics. Others are longer term, reflecting the need to continue reassessing the needs of U.S. high school mathematics students.

Implementation Challenges

In the following sections, we discuss several major challenges related to implementation, including the need for professional development, relevant instructional materials, and effective assessment systems.

Professional Development

Professional development needs are extensive if the CCSSM is to be implemented successfully. Many teachers have had limited experience with mathematics that focuses more on reasoning and problem solving than on carrying out procedures (Borasi and Fonzi 2002). They may also have had limited exposure to instructional practices that help students meet the Standards for Mathematical Practice. In addition, some of the content foci (such as the emphasis on geometric transformations and the substantial work with statistics) may also be unfamiliar. Thus, teachers need professional development in the areas of mathematical content, instructional practices, curriculum issues, and assessment. Furthermore, this development must be undertaken for both in-service and preservice teachers, which in particular means that higher education institutions must be proactively involved. Unless a significant effort is made to help teachers understand both the intent and particulars of the CCSSM better, it will be difficult for teachers to adapt to the new requirements and for students to make the intended gains in understanding and achievement.

Instructional Materials

Given the central role of the textbook in the decision making of many teachers (Schmidt, McKnight, and Raizen 1997), instructional materials that support the vision of the CCSSM and meet its standards are essential. Simply reorganizing an existing curriculum or textbook by cut-and-paste and making other minor adjustments will not suffice. The CCSSM's content may require some rethinking of not only what content a new or modified curriculum includes, but also, even more difficult, how that new curriculum conceptualizes, organizes, and teaches its content in order to develop the required deeper understanding. Furthermore, instructional materials must support developing the Standards of Mathematical Practice across the curriculum, not just as additional topics or problems tacked onto the end of homework assignments. The mathematical practices must be an inherent part of the content development. This relates to the problem-solving theme discussed earlier, in that we must find or create instructional materials and practices that allow us to teach *through* the practices identified in the Standards of Mathematical Practice, not just teach about these practices. A curriculum analysis tool developed by Bush et al. (2011) gives teachers and other decision-makers important information about the effectiveness of particular materials toward meeting the CCSSM recommendations, including whether the materials address the mathematical practices productively.

Curriculum Sequence

Besides the challenges related to the substance of the curriculum recommended in the CCSSM, challenges also involve the sequence of the curriculum. What sequences of courses will meet the CCSSM standards effectively? For example, a number of innovative curricula appearing over the past decade or more—Core-Plus Mathematics, the Interactive Mathematics Program, and others—emphasize problem solving and reasoning and have a broad view of the high school mathematics curriculum (cf. Hirsch, 2007). These characteristics make them generally consistent with the CCSSM and thereby useful to meet the challenge of instructional materials just discussed. However, these curricula are integrated, making deliberate connections among algebra, geometry, and other strands each year. Such an integrated approach is common outside the United States, and *Appendix A* of the CCSSM (CCSSI 2010b) offers an integrated pathway as an explicit option. Integrated approaches, however, are less common in the United States, Moreover, no consensus on the precise configuration of an integrated approach exists. So what courses and pathways can states and districts design that support an integrated approach? If the more common, nonintegrated, single-subject sequence is maintained, how can we sequence and configure those courses to meet the CCSSM recommendations for all students, especially regarding new content or new expectations in existing content? Overcoming the lack of definition and consistency in the curriculum pathways that states will employ to meet the CCSSM standards for high school mathematics will be a major challenge.

Assessment Systems

Our premise is that the accountability that the CCSSM provides may be a major spur for positive change in high school mathematics. However, that will be true only if the developed accountability systems incorporate significant attention to the mathematical practices and encompass the deeper understanding of mathematics set forth in the CCSSM. Otherwise, teachers will have no impetus to take a different approach to the curriculum. This is a particularly thorny problem since most high-stakes assessments to date do not strongly address either the mathematical practices or deep understanding and application of the content. Although a commitment exists among the assessment consortia to develop models that measure the effectiveness of the CCSSM at a deep level, as with the organization of the high school mathematics curriculum, each state is free to make its own decisions about what assessment system it adopts. The lack of consensus on pathways for achieving the high school standards only adds to the challenge of creating new, powerful assessments that are usable across the states.

Challenges of Continual Improvement

Although the CCSSM has achieved broad acceptance, it is important to see the CCSSM as a point in the journey to improve high school mathematics education, not the final destination. As the document's writers themselves have insisted, the CCSSM will need continuing revision as we learn more about how these standards play out across the country (cf. McCallum 2010). We identify two particular areas that will need continuing attention and reflection.

Content

Debate will inevitably follow any set of content recommendations for school mathematics. Is it the right content, at the right depth, and at the right time for all students? The CCSSM laudably recommends broad, deep content. For example, we have finally reached consensus on including statistics as a core component of a mathematical sciences curriculum for all high school students. However, the CCSSM does not reflect a forward-looking perspective on other potential core topics related to the information-dense digital age in which we live. For example, fundamental aspects of life in a modern democratic society include voting, the Internet, and networks. It seems reasonable that students need some level of quantitative literacy in these areas; examples might include understanding some basic mathematics of ranked-choice voting, some elementary number theory and logic as used in Internet security and searching, and basic concepts and methods of graph theory (vertex-edge graphs) as they relate to many and varied problems dealing with networks in society, science, and everyday life. The CCSSM neglects many such topics that are of both mathematical and practical interest.

Another aspect of the content needing continued attention relates to the CCSSM's "+" content category. For example, why does the CCSSM classify all work with matrices in the "+" category, whereas it deems using polynomial identities to generate Pythagorean triples core for all students? One might easily argue that matrices have a much more pervasive role in their use as mathematical models to organize and manipulate data.

Standards, not Standardization

Although this chapter's premise, as well as the CCSSM's premise, is that a clear vision and strong standards are essential for improving mathematics education, a danger arises when standards become rigid standardization. This can lead to weakened content expectations, in order to make those expectations attainable for all students. Furthermore, standardization can inhibit educational innovation with its pressure to meet what often become rigid requirements. Figuring out how to have high standards without the limitations of rigid standardization will continue to be a major challenge that needs vigilant attention.

Conclusion

As we move into the era of the Common Core State Standards for high school mathematics, we face a tremendous opportunity to reassess what we are doing. However, unless we continue the dialogue about how best to improve high school mathematics education—across the nation, as well as in particular states, local districts and schools, and as a profession—we will all too likely come up short, undertaking a proverbial shuffling of deck chairs on the *Titanic* instead of making the dramatic moves needed to change the ship's direction. NCTM can provide some of the immediate resources needed to begin a deliberate dialogue about high school mathematics, including this volume, the High School Mathematics Project (cf. Shaughnessy 2010), and a joint action plan forged with three other major mathematics education organizations (Joint Task Force on Common Core State Standards 2010), which will address some of the issues related to implementation. Ensuring that teachers have the knowledge, resources, and support they need to enact this new vision will take concerted effort among *all* stakeholders in high school mathematics. We must reconsider the system as a whole—curriculum, instruction, and assessment—for its alignment with that vision. We must also realize that this is a process of *continual* improvement that will require reassessing and modifying the CCSSM over the coming years.

The future success, not just of our students but also of our nation, depends on our pursuing this dialogue and then acting to produce meaningful change.

References

Achieve, Inc. *High School Model Three-Year Integrated Course Sequence*. Washington, D.C.: Achieve, Inc., 2007a.

———. *High School Model Three-Year Traditional Plus Course Sequence*. Washington, D.C.: Achieve, Inc., 2007b.

———. *Secondary Mathematics Benchmarks Progressions*. Washington, D.C.: Achieve, Inc., 2007c.

Aldeman, Chad. *College- and Career-Ready: Using Outcomes Data to Hold High Schools Accountable for Student Success*. Washington, D.C.: Education Sector, 2010. http://www.educationsector.org.

American Diploma Project (ADP). *Ready or Not: Creating a High School Diploma That Counts*. Washington, D.C.: Achieve, Inc., 2004.

Borasi, Raffaella, and Judith Fonzi. *Professional Development That Supports School Mathematics Reform*, Foundations Monograph, Vol. 3. Arlington, Va.: National Science Foundation, 2002.

Bush, William et al. *Curriculum Analysis Tool for the Common Core State Standards in Mathematics*, 2011.

College Board. *Standards for College Success: Mathematics and Statistics.* New York: College Board, 2006.

———. *Standards for College Success: Mathematics and Statistics—Adapted for Integrated Curricula.* New York: College Board, 2007.

Committee on Prospering in the Global Economy of the 21st Century. *Rising Above the Gathering Storm: Energizing and Employing America for a Brighter Economic Future.* Washington, D.C.: National Academies Press, 2007.

Common Core State Standards Initiative (CCSSI). *Appendix A: Designing High School Mathematics Courses Based On the Common Core State Standards.* Washington, D.C.: National Governors Association Center for Best Practices and the Council of Chief State School Officers, 2010b. http://www.corestandards.org.

———. *Common Core State Standards for Mathematics.* Washington, D.C.: National Governors Association Center for Best Practices and the Council of Chief State School Officers, 2010a. http://www.corestandards.org.

Franklin, Christine, Gary Kader, Denise Mewborn, Jerry Moreno, Roxy Peck, Mike Perry, and Richard Scheaffer. *Guidelines for Assessment and Instruction in Statistics Education (GAISE) Report: A Pre-K–12 Curriculum Framework.* Alexandria, Va.: American Statistical Association, 2007.

Great Schools. "High School Exit Exams: Issues to Consider," 2010. http://www.greatschools.org.

Hirsch, Christian R., ed. *Perspectives on the Design and Development of School Mathematics Curricula.* Reston, Va.: National Council of Teachers of Mathematics, 2007.

Hirsch, Christian, Dana Cox, Lisa Kasmer, Sandy Madden, and Diane Moore. "Some Common Themes and Notable Differences across Recent National Mathematics Curriculum Documents." In *K–12 Mathematics: What Should Students Learn and When Should They Learn It?—Conference Highlights,* edited by the Center for the Study of Mathematics Curriculum (CSMC), pp. 40–51. Columbia, Mo.: CSMC, 2007.

Joint Task Force on the Common Core State Standards. *Report of the Joint Task Force on Common Core State Standards.* Association of Mathematics Teacher Educators, Association of State Supervisors of Mathematics, National Council of Supervisors of Mathematics (NCSM), and National Council of Teachers of Mathematics, 2010. http://nctm.org/uploadedFiles/About_NCTM/President/Messages/Shaughnessy/2010_1104_PresMess_B.pdf.

Lockhart, Paul. "A Mathematician's Lament." In "Devlin's Angle," *MAA Online,* March 2008. http://www.maa.org/devlin/LockhartsLament.pdf.

Mathematical Association of America (MAA). *Curriculum Foundations Project: Voices of the Partner Disciplines.* Curriculum Renewal across the First Two Years (CRAFTY), subcommittee of the Committee on the Undergraduate Program in Mathematics (CUPM). Washington, D.C.: MAA, 2004.

———. *Algebra: Gateway to a Technological Future.* Washington, D.C.: MAA, 2007a. Summary at http://www.maa.org/algebra-report/index.html.

———. *CRAFTY Guidelines for College Algebra.* Curriculum Renewal across the First Two Years (CRAFTY), subcommittee of the Committee on the Undergraduate Program in Mathematics (CUPM). Washington, D.C.: MAA, 2007b.

McCallum, William. Remarks at the CBMS Forum on Content-Based Professional Development for Teachers of Mathematics, Reston, Va., October 10–12, 2010.

Members of the 2005 "Rising Above the Gathering Storm" Committee. *Rising Above the Gathering Storm, Revisited: Rapidly Approaching Category 5.* Washington, D.C.: National Academies Press, 2010.

Murray, Vicki E. *The High Price of Failure in California: How Inadequate Education Costs Schools, Students, and Society.* San Francisco: Pacific Research Institute, 2009. http://www.pacificresearch.org.

National Center for Education Statistics (NCES). *A Profile of the American High School Senior in 2004.* U.S. Department of Education, Institute of Education Sciences. Washington, D.C.: NCES, 2005.

———. *The Nation's Report Card: Long-Term Trends 2008.* Washington, D.C.: Institute of Education Sciences, U.S. Department of Education, 2009.

———. *The Nation's Report Card: Grade 12 Reading and Mathematics 2009 National and Pilot State Results.* Washington, D.C.: Institute of Education Sciences, U.S. Department of Education, 2010.

National Commission on the High School Senior Year (NCHSSY). *Raising Our Sights: No High School Senior Left Behind.* Washington, D.C.: U.S. Department of Education, 2001.

National Council of Teachers of Mathematics (NCTM). *An Agenda for Action.* Reston, Va.: NCTM, 1980.

———. *Curriculum and Evaluation Standards for School Mathematics.* Reston, Va.: NCTM, 1989.

———. *Principles and Standards for School Mathematics.* Reston, Va.: NCTM, 2000.

———. *Focus in High School Mathematics: Reasoning and Sense Making.* Reston, Va.: NCTM, 2009.

———. *Making It Happen: A Guide to Interpreting and Implementing the Common Core State Standards for Mathematics.* Reston, Va.: NCTM, 2011.

Partnership for 21st Century Skills. *Are They Really Ready to Work? Employers' Perspectives on the Basic Knowledge and Applied Skills of New Entrants to the 21st Century U.S. Workforce.* Report on results of a nationwide poll in April and May, 2006.

———. *Beyond the Three Rs: Voter Attitudes toward 21st Century Skills.* Report on results of a nationwide poll in September, 2007.

Programme for International Student Assessment (PISA). *PISA 2009 Results: What Students Know and Can Do: Student Performance in Reading, Mathematics and Science,* Vol. 1. Paris: Organization for Economic Cooperation and Development, 2010.

Schmidt, William H., Curtis C. McKnight, and Senta A. Raizen. *A Splintered Vision: An Investigation of U.S. Science and Mathematics Education.* East Lansing, Mich.: U.S. National Research Center for the Third International Mathematics and Science Study, Michigan State University, 1997.

Schoen, Harold L., ed. *Teaching Mathematics through Problem Solving, Grades 6–12.* Reston, Va.: National Council of Teachers of Mathematics, 2003.

Shaughnessy, J. Michael. "NCTM Mobilizes CCSS Implementation Support and Continues Its Focus on Reasoning and Sense Making," September 2010. http://nctm.org/about/content.aspx?id=26483.

Singapore Ministry of Education. *Secondary Mathematics Syllabuses.* 2006.

Steen, Lynn Arthur. "Facing Facts: Achieving Balance in High School Mathematics." *Mathematics Teacher* 100 (December 2006): 86–95.

U.S. Department of Education. "U.S. Secretary of Education Duncan Announces Winners of Competition to Improve Student Assessments," September 2, 2010. http://www.ed.gov/news/press-releases/us-education-secretary-duncan-announces-winners-competition-improve-student-asse.

Section III
Building Curriculum Coherence: Case Studies

Curriculum Issues in an Era of Common Core State Standards for Mathematics

Introduction

The *Common Core State Standards for Mathematics* (CCSSM; CCSSI 2010) represents a coordinated effort to bring greater focus and coherence to school mathematics programs in grades K–12. The CCSSM articulates the mathematics that students are expected to study in each grade, K–8, and across the high school grades. The standards do not dictate curriculum organization or teaching style. Yet, we know that "effective mathematics teaching requires understanding what students know and need to learn and then challenging and supporting them to learn it well" (Teaching Principle, NCTM 2000, p. 16).

District-provided curriculum materials (e.g., textbooks) are the primary influence on how teachers organize their mathematics instruction. However, effective teachers do not simply use the textbook as written; they adapt it to their students' needs, the CCSSM, possibly additional state or district mandates, and their own beliefs about teaching and learning mathematics. In Section III we focus on teachers' interaction with curriculum materials. The following three case studies examine teachers' use of curriculum materials through a focus on curriculum vision and coherence. Each chapter considers these questions:

- How do we learn to use and adapt curriculum materials in ways that allow us to meet our students' needs *and* address the multiple standards for which we and our students are responsible?
- What issues and questions might we consider in making decisions about our use of curriculum materials?

The team of Michelle Cirillo, Corey Drake, and Beth Herbel-Eisenmann prepared this section's three chapters. The first observes Nancy, a second- and third-grade teacher, as she builds a coherent view of her curriculum. The second presents the case of Stacey, a seventh- and eighth-grade teacher, as she plans with and enacts her curriculum vision while maintaining curriculum coherence. The third follows Matt, a high school geometry teacher, as he works to implement his curriculum vision with a focus on coherence.

As you read these three chapters, think about their implications for building CCSSM-oriented curriculum coherence in your school district, school, and classroom teaching.

Questions for Reflection and Collective Discussion

1. What did you find particularly interesting or challenging in your reading of one or more of the chapters in this section? Be prepared to discuss those ideas with your colleagues and, where possible, identify possible ways to improve practice.
2. How does the set of chapters help you examine (*a*) your vision for what students can and should learn in mathematics and (*b*) how your curriculum materials support or do not support this vision?
3. From your perspective, what is the role of the curriculum materials your school chooses for mathematics?
4. How do you approach developing your plans for teaching mathematics each day? In what ways do you and your colleagues work together to improve students' opportunities to learn?

5. What does the statement "the relationship that teachers have with their curriculum often resembles, at best, 'an arranged marriage'" mean to you? How would you characterize your relationship with your school curriculum?
6. The chapters challenge you to think about "curriculum vision." What does this mean to you?
7. How could you and your colleagues begin to work toward developing a coherent and shared curriculum vision to guide your teaching and your goals?
8. In Matt's case study, you see a young teacher establishing his stance toward teaching. How does Matt's explication of his approach to engaging students resonate with your own practice?
9. What would be some ways to begin changing the discourse practices in your classroom so that students are learning to think, reason, and engage in productive mathematical discussions?
10. How do your state standards for elementary school mathematics compare with the corresponding standards in the *Common Core State Standards for Mathematics* (CCSSM) (http://www.corestandards.org)? What major differences will your school or district need to address as it works toward curriculum alignment with the CCSSM? Working with your colleagues, make a chart summarizing options for addressing identified differences.
11. How do your state standards for middle school mathematics compare with the corresponding CCSSM standards (http://www.corestandards.org)? What major differences will your school or district need to address as it works toward curriculum alignment with the CCSSM? Working with your colleagues, make a chart summarizing options for addressing identified differences.
12. How are your state standards for high school mathematics similar to, and different from, the corresponding CCSSM standards? What major differences will your school or district need to address as it works toward curriculum alignment with the CCSSM? Working with your colleagues, make a chart summarizing options for addressing identified differences.
13. Curriculum materials (e.g., textbooks) determine strongly what students have the opportunity to learn, what they learn, and how they learn it. Curriculum materials that truly align with the CCSSM content standards and mathematical practices will be important in implementing the CCSSM effectively. Become familiar with the Council of Chief State School Officers (CCSSO) document, *CCSSO Mathematics Curriculum Analysis Tools* (Bush and Ronau 2011).

    a. How might your district use these tools in selecting curriculum materials that align with the CCSSM?

    b. How will your district judge the coherence of curriculum materials that are found to align well with the CCSSM?

14. Select an appropriate mathematical idea from the CCSSM, not currently part of your curriculum, and work with colleagues to design and teach a lesson on that idea, paying careful attention to what discourse patterns you are trying to promote. Reflect and report on the experience from the perspectives of teaching, students' engagement, and learning.

References

Bush, William S., and Robert N. Ronau, eds. *Council of Chief State School Officers (CCSSO) Mathematics Curriculum Analysis Tools.* Washington, D.C.: CCSSO, 2011.

Common Core State Standards Initiative. *Common Core State Standards for Mathematics.* Washington, D.C.: National Governors Association Center for Best Practices and the Council of Chief State School Officers, 2010. http://www.corestandards.org.

National Council of Teachers of Mathematics (NCTM). *Principles and Standards for School Mathematics.* Reston, Va.: NCTM, 2000.

Chapter 5

Using Curriculum to Build on Children's Thinking

Corey Drake
Michelle Cirillo
Beth Herbel-Eisenmann

THE RELATIONSHIP that teachers have with their curriculum materials often resembles, at best, "an arranged marriage" (The El Barrio–Hunter College PDS Partnership Writing Collective 2009). In some cases, this arrangement is compatible, and teachers feel comfortable using the materials as written; but in many cases, the arrangement does not work well. At the same time, in an era of rigorous assessment and accountability systems, teachers often face long lists of mathematics topics or learning expectations that must be addressed throughout the school year (NCTM 2006). More specifically, classroom teachers are expected to attend to national standards and sometimes additionally state standards and district standards, using mass-produced curriculum materials that were written neither for unique contexts nor for individual students.

Despite these potential incompatibilities, mathematics teachers tend to rely heavily on commercial curriculum materials as their primary tool for teaching mathematics (Grouws, Smith, and Sztajn 2004).

Curriculum Vision and Coherence

Researchers support the assertion that to be successful, teachers must develop a curriculum vision. For example, "well-prepared teachers have developed a sense of 'where they are going' and how they and their students are going to get there" (Darling-Hammond et al. 2005, p. 177). We propose that a teacher with a well-defined curriculum vision has a deep understanding of (1) "big mathematical ideas" or goals of the curriculum; (2) multiple trajectories that students might take to reach those goals; and (3) strategies for using, adapting, supplementing, or creating curriculum materials in order to support students in reaching those goals. An important component of curriculum vision is

---

Adapted from Drake, Corey, Michelle Cirillo, and Beth Herbel-Eisenmann. "Using Curriculum to Build on Children's Thinking." *Teaching Children Mathematics* 16 (August 2009): 49–54.

This research was supported, in part, by the National Science Foundation under grant no. 0347906 (Beth Herbel-Eisenmann, principal investigator [PI]) and grant no. 0643497 (Corey Drake, PI). Any opinions, findings, conclusions, or recommendations expressed are those of the authors and do not necessarily reflect the views of the National Science Foundation. We are grateful to Nancy for opening her classroom to us during this study.

curriculum coherence—mathematical connections that can be made by both students and teachers—within and across lessons.

In the sections that follow, we offer suggestions to help teachers develop, plan with, and enact a well-defined curriculum vision. We also present data from interviews with Nancy, who teaches second and third grade, in order to provide examples of the kinds of questions that elementary teachers might consider as they think through curricular issues.

Developing the Vision

The Common Core State Standards, like *Principles and Standards*, advocates the idea that math curriculum be coherent: "A curriculum is more than a collection of activities: It must be coherent, focused on important mathematics, and well articulated across the grades" (NCTM 2000, p. 14). We believe that this kind of coherence can be achieved through the development and enactment of curriculum vision. Recognizing that different mathematical content ought to "receive different emphases across the grade bands" can help teachers think about the "big ideas" that students ought to learn throughout the year.

Other questions that teachers might ask themselves concern the focus and coherence of the curriculum, learning trajectories of specific topics (cf. Confrey, Maloney, and Nguyen 2010) connections within and across mathematical topics, and the relationship between curriculum content and pedagogy, or how the content might be taught. Finally, in developing their curriculum vision, teachers will want to reflect on and include their own beliefs as well as consider specific questions regarding their particular school and district contexts. In the next section, we use data from interviews to illustrate how Nancy, an elementary teacher, explored such questions as she developed a curriculum vision for her mathematics instruction.

Nancy's Vision

Although every teacher has his or her own process for developing a curriculum vision, we find that, in general, elementary teachers develop strong curriculum vision by considering, studying, and using one or more of three important resources: (1) district standards or grade-level expectations, (2) their students' mathematical thinking, and (3) mathematics curriculum materials. Teachers often describe using the standards and curriculum materials to develop their understanding of the mathematical goals for their grade levels, as well as for the big mathematical ideas to be learned during the year. Teachers' understanding and assessments of their own students' mathematical thinking are a third resource that can then be used to assist teachers as they make instructional decisions on a daily and weekly basis in order to accomplish their goals.

In an interview related to lesson planning and instructional decision-making processes, Nancy discussed how she developed and refined her curriculum vision. She described her experiences with mathematics curriculum materials that focused her attention on questioning and understanding children's problem-solving strategies; her participation in Cognitively Guided Instruction (CGI) (Carpenter et al. 1999); and her work on the district curriculum map (an outline of standards, goals, and objectives for particular grade levels). In particular, these experiences strengthened and clarified her vision for student learning over the course of the school year:

CGI has really helped me understand how my students are thinking about math. So that has heightened my level of awareness.... And so that's really raised my understanding of how kids think about numbers.... So how do I get the kid who is taking out three cubes, taking out eight cubes and then re-counting them all.... How do I get her to say *eight*, either *two more* [than eight] *is ten* [and] *one more* [than ten] *is eleven*, or *eight, nine, ten, eleven* when it's like a three plus eight problem?

Nancy's experiences with CGI helped her better understand how her students think about mathematics. This new understanding has, in turn, led her to refine the mathematical goals for her students. In particular, a clear part of her vision for her students is that they will develop and use efficient strategies for solving multidigit addition and subtraction problems. As she indicated, this vision also includes an understanding of the steps, or strategies, necessary for bridging where her students are currently and where she wants to take them.

> **Developing Curriculum Vision and Coherence**
>
> Teachers will want to consider and include both their own beliefs and specific questions regarding their particular school and district contexts.
>
> 1. What are the "big ideas" (important mathematical content) that I want students to learn?
> 2. Is my curriculum one that is coherent and focused rather than a collection of activities or topics? How do I create coherence or connections across these ideas?
> 3. What are the ways through which I want my students to acquire and use this content knowledge? In other words, how can I teach this content while still addressing the important mathematical processes?
> 4. What do I see as the balance among skills, procedures, and concepts?
> 5. How do my own knowledge, beliefs, and experiences fit into this vision?
> 6. How do my individual students' needs influence my curriculum vision?
> 7. How do I consider particular features of high-quality mathematics education (i.e., NCTM's principles) as guides and tools for making decisions about curriculum?

Nancy was also careful to point out that before she participated in CGI workshops, her many years of experience with NSF-funded, Standards-based curriculum materials helped her develop a foundation for her curriculum vision. She believes that all teachers should have the opportunity to learn from curriculum materials and to develop curricular understandings that encompass not only their own grade level expectations (horizontal curriculum) but also the grade levels above and below (vertical curriculum) in order to respond to the needs of students who might be working mathematically above or below grade level. She explained,

> What are you going to do with kids who aren't [performing] at [the] second-grade math level and [other] kids who are ready to move on? ... That flexibility to move beyond or in between those curriculum maps or expectations ... is important.

Nancy's curriculum vision exemplifies the way in which teachers can understand mathematical content and goals for their particular grade levels as well as connect those goals to the needs of their individual students. Because developing a coherent curriculum vision is only half the curriculum vision equation, specific questions can assist teachers in planning with and enacting their curriculum vision.

Planning and Enacting a Curriculum Vision

Teachers who are familiar with their resources and work at learning about and assessing their students' thinking can pick and choose among resources when enacting a curriculum vision.

1. What is the implicit curriculum vision put forth by the authors of my curriculum materials?
2. In what ways is my curriculum vision aligned with or different from that of my curriculum materials?
3. How do the standards or objectives of my state and district fit into this vision?
4. How do I take into account both the horizontal curriculum (within my course) and also the vertical curriculum (across the grades band)?
5. How can I use my curriculum materials as a tool to achieve my curriculum vision?
6. Do I need to adapt, reorganize, or delete elements of the curriculum materials? If so, how?
7. Do I need to supplement the curriculum materials with additional ideas or activities? If so, how?
8. How does my curriculum vision impact the ways that I interact with my students?
9. How do my individual students and their communication about their mathematical thinking influence my curriculum vision?

Planning with and Enacting the Vision

One important step in planning to enact a curriculum vision is to "look at your curriculum and begin to organize it into more coherent instruction" (Mirra 2008, p. 7). That is, examining the existing state and district curriculum guidelines in conjunction with available curriculum materials can provide a starting point for planning with and enacting a curriculum vision. When looking at the curriculum materials, teachers might identify the curriculum vision (either implicit or explicit) put forth by the curriculum's authors. Teachers can then consider how their curriculum vision is aligned with or different from that of the curriculum materials and how their state or district standards fit into this vision. In order to achieve coherence across grade levels, teachers should also consider the mathematics that students have experienced prior to their current course and the mathematics that they will need to know to be prepared for subsequent courses.

Next, looking carefully at their curriculum materials, teachers will need to make decisions about using and adapting the materials.

Finally, teachers must consider how their curriculum vision will influence the enactment of their lessons and their interactions with students during instruction. In the next section, we look again to Nancy to provide examples of how to plan with and enact a curriculum vision.

Nancy Enacts Her Vision

Just as Nancy's experiences with curriculum materials, CGI, and the district's curriculum mapping process were the basis for developing her curriculum vision, her day-to-day enactment of her curriculum vision also draws on these various resources and experiences. For example, Nancy described the enactment of a recent lesson and the ways in which she was able to draw on a variety of resources to design instruction:

> I just really look at what I see [my students] doing, and any misconceptions that they have about numbers that they're working with. And I think a great example is today.... I was noticing that the kids weren't understanding or using doubles to help them add or subtract. And so ... we read a literature book [*Two of Everything*, Hong 1993] ... and then they wrote their own [doubles] problems ... and then I let them choose their numbers and ... that told me a lot about them and their understanding of number. So, really that's kind of my basis for the kids is really watching them, questioning them, reviewing what they're doing and having a process time at the end of my math period where we talk about it because I find ... then that helps me guide the next day's discussion.

Nancy went on to describe how, more generally, her well-defined curriculum vision, supported by extensive experience with curriculum materials and CGI, guides her decisions about the kinds of resources she chooses to use and how she chooses to use them:

> You know, there's times that I'll pull a specific lesson because I'm very familiar [since] I have worked with the [curriculum materials for a while]. I'm used to their lessons, and I knew there was, like today, a good doubling lesson, so I went to that. Or I know there's a great game to help kids jump by a hundred, tens, and ones. So I'll pull that out and bring that in. I guess it's my familiarity with [the materials] that has helped me be able to do that.

As Nancy explained, when enacting her curriculum vision, she draws on a variety of resources. She can pick and choose among these resources not only because she has gained familiarity with the resources themselves but also because she works to learn about and assess her students' thinking. Nancy's curriculum vision, which includes mathematical goals for her students and understanding the ways in which they might meet those goals, guides her instructional decisions and ensures that the mathematics instruction is coherent for students.

Conclusion

Nancy has been able to use her experiences with a curriculum program that is focused on questioning and problem solving, as well as her knowledge of her students and mathematics, to develop a curriculum vision that allows her to enact a coherent mathematics course that continually assesses and builds on her students' mathematical thinking. As Nancy participates in new professional experiences, such as CGI training and curriculum mapping, she continues to reflect on and develop her teaching, working to align her curriculum vision with her practice.

Many continually changing factors influence classroom teachers' practices: national, state, and district standards; new textbooks; technological advancements; research about how students learn mathematics; and so forth. One constant remains, however: teachers' daily interactions with their students. Ultimately, the teacher's job is to do what is best for his or her students. Just as Nancy carefully reflected on a curriculum vision, teachers can use curriculum materials and other instructional resources in ways that adapt to their students in their own particular contexts. In fact, we believe that teachers are the only ones who *can* do this kind of work because they are the ones who know their students, their school, and their community. However, teachers need not do this work alone. Colleagues within a school or district can embark on a collaborative journey of defining and developing a curriculum vision.

References

Carpenter, Thomas P., Elizabeth Fennema, Megan L. Franke, Linda Levi, and Susan B. Empson. *Children's Mathematics: Cognitively Guided Instruction*. Portsmouth, N.H.: Heinemann, 1999.

Confrey, Jere, Alan P. Maloney, and Kenny H. Nguyen. *Learning Trajectory Display of the Common Core State Standards for Mathematics, Grades K–5*. New York: Wireless Generation, 2010.

Darling-Hammond, Linda, James Banks, Karen Zumwalt, Louis Gomez, Miriam Gamoran Sherin, Jacqueline Griesdorn, and Lou-Ellen Finn. "Developing a Curricular Vision for Teaching." In *Preparing Teachers for a Changing World: What Teachers Should Learn and Be Able to Do*, edited by Linda Darling-Hammond and John Bransford. San Francisco: Jossey-Bass, 2005.

The El Barrio–Hunter College PDS Partnership Writing Collective. "On the Unique Relationship between Teacher Research and Commercial Mathematics Curriculum Development." In *Mathematics Teachers at Work: Connecting Curriculum Materials and Classroom Instruction*, edited by Janine T. Remillard, Beth A. Herbel-Eisenmann, and Gwendolyn M. Lloyd, pp. 118–33. New York: Routledge, 2009.

Grouws, Douglas A., Margaret S. Smith, and Paola Sztajn. "The Preparation and Teaching Practices of United States Mathematics Teachers: Grades 4 and 8." In *Results and Interpretations of the 1990 through 2000 Mathematics Assessments of the National Assessment of Educational Progress*, edited by Peter Kloosterman and Frank K. Lester, Jr. Reston, Va.: National Council of Teachers of Mathematics, 2004.

Hong, Lily Toy. *Two of Everything: A Chinese Folktale*. Morton Grove, Ill.: Albert Whitman & Company, 1993.

Mirra, Amy. *Focus in Grades 3–5: Teaching with Curriculum Focal Points*. Reston, Va.: National Council of Teachers of Mathematics, 2008.

National Council of Teachers of Mathematics (NCTM). *Principles and Standards for School Mathematics*. Reston, Va.: NCTM, 2000.

———. *Curriculum Focal Points for Prekindergarten through Grade 8 Mathematics: A Quest for Coherence*. Reston, Va.: NCTM, 2006.

Chapter 6

Using Curriculum to Focus on Understanding

Michelle Cirillo
Beth Herbel-Eisenmann
Corey Drake

THE RELATIONSHIP that teachers have with their curriculum materials often resembles, at best, "an arranged marriage" (The El Barrio-Hunter College PDS Partnership Writing Collective 2009). In some cases, this arrangement is compatible, and teachers feel comfortable using the materials as written. In many cases, it is not. At the same time, in an era of Common Core State Standards for Mathematics (CCSSM) (Common Core State Standards Initiative 2010) linked to rigorous assessment and accountability systems, teachers often face long lists of mathematics topics or learning expectations that need to be addressed throughout the school year. More specifically, today classroom teachers are expected to address the CCSSM and/or state standards using mass-produced curriculum materials that were not written for their unique contexts or for their individual students.

Despite these potential incompatibilities, mathematics teachers tend to rely heavily on commercial curriculum materials as their primary tool for teaching mathematics (Grouws, Smith, and Sztajn 2004). Two questions come to mind:

1. How do we learn to use and adapt impersonal curriculum materials in ways that allow us to meet our students' needs and address the multiple standards and objectives for which we and our students are responsible?

2. What kinds of issues and questions might we consider in making decisions about our use of curriculum materials?

In this chapter, we describe *curriculum vision;* it includes an understanding of the mathematics that students must learn and a coherent trajectory (or set of trajectories)

Adapted from Cirillo, Michelle, Beth Herbel-Eisenmann, and Corey Drake. "Using Curriculum to Focus on Understanding." *Mathematics Teaching in the Middle School* 15 (August 2009): 51–56.

This research was supported, in part, by the National Science Foundation under grant no. 0347906 (Beth Herbel-Eisenmann, principal investigator [PI]) and grant no. 0643497 (Corey Drake, PI). Any opinions, findings, conclusions, or recommendations expressed in this material are those of the authors and do not necessarily reflect the views of the NSF.

for learning that mathematics. We propose that teachers who manage the complex task of monitoring students' mathematical needs as well as standards and objectives have a well-defined curriculum vision. We offer suggestions for ways that teachers can reflect on and articulate this view.

Curriculum Vision and Coherence

Researchers have suggested that successful teachers need to develop a "curriculum vision." For example, Darling-Hammond et al. (2005) claimed that "well-prepared teachers have developed a sense of 'where they are going' and how they and their students are going to get there" (p. 177). We propose that a teacher with a well-defined curriculum vision has a deep understanding of—

1. "big mathematical ideas," or goals of the curriculum;
2. the multiple trajectories that students might take to reach those goals; and
3. strategies for using, adapting, supplementing, or creating curriculum materials to support students in reaching those goals.

An important component of curriculum vision is curriculum coherence—mathematical connections that can be made by both students and teachers—within and across lessons.

In the sections that follow, we offer suggestions and questions to help teachers develop, plan, and enact their curriculum vision. We also present data from interviews with Stacey, a middle school mathematics teacher, to show the kinds of issues and questions that teachers might consider as they think through curricular issues.

Developing a Curriculum Vision

The idea that a mathematics curriculum should be coherent is advocated in the CCSSM as it is in *Principles and Standards for School Mathematics*: "A curriculum is more than a collection of activities: it must be coherent, focused on important mathematics, and well articulated across the grades" (NCTM 2000, p. 14). We believe that this kind of coherence can be achieved through the development and enactment of curriculum vision.

Recognizing that different mathematical content ought to receive different emphases across the grade bands can help teachers think about the big ideas, their learning trajectories (cf. Confrey, Maloney, and Nguyen 2010), and what students should focus on in a given year. Attention also should be given to the focus and coherence of the curriculum, connections within and across mathematical topics, and the relationship between curriculum content and pedagogy. Teachers may consider how the content will be developed with students.

Finally, in developing their curriculum vision, teachers will want to reflect on and include their own beliefs as well as consider them in light of their particular school and district. We provide specific questions for teachers to consider in figure 6.1.

> - What are the "big ideas" (important mathematical content) that students ought to learn?
> - Is my curriculum coherent and focused, rather than a collection of activities or topics? How do I create coherence or connections across these ideas?
> - What are the ways through which I want my students to acquire and use this content knowledge? In other words, how can I teach this content, while still addressing the important mathematical processes?
> - What do I see as the balance among skills, procedures, and concepts?
> - How do my own knowledge, beliefs, and experiences fit into this vision?
> - How do my individual students' needs influence my curriculum vision?
> - How do I consider particular features of high-quality mathematics education (e.g., NCTM's principles for school mathematics) as guides and tools for making decisions about curriculum?

Fig. 6.1. Questions to consider in developing curriculum vision and coherence

How Stacey Developed a Curriculum Vision

Stacey's early teaching reflected how she was taught:

> My mathematics instruction for the first several years of my teaching was traditional in that I taught the concepts and algorithms and expected students to practice them repeatedly. This is how I learned mathematics, and I had never thought of other possibilities.

However, after reading NCTM's *Standards* documents (1989, 2000) and attending conferences, Stacey's thinking about mathematics teaching began to change. These professional activities caused Stacey to reflect on her own practice and realize new possibilities for teaching mathematics in more meaningful ways.

As her goals changed, she became interested in using an NSF-funded, Standards-based curriculum program to further her vision that "students need to learn mathematics conceptually with a focus on understanding." When asked to describe her ideal teaching situation, Stacey mentioned these curriculum materials:

> It would be *ideal* to be able to use the type of materials that fit my philosophy,

and I do. I haven't always and now I do. Ideal to me is to be able to use a more investigative type of curriculum and be able to supplement with skill work, rather than the other way around.

Although Stacey was satisfied with the materials she was using, she also found that enacting the lessons in those curriculum materials required a clear sense of both her role and that of her students. She said that students in her classroom should take charge of their own learning: "It's not up to me to open up the think tank and pour it in. It's [their] job to work with me and figure it out." Although she expected her students to be active learners, she also believed that she, herself, played an important role:

> My role is to make the curriculum come alive. My role is to help [my students] see mathematics in a way that is meaningful, that is useful ... and to facilitate that learning.... My role is to assess that learning and figure out when it's happening, when it's not happening, and how I can get it to happen.

Stacey found this new role to be challenging. Rather than directly teaching the material as she had done in the past, she now believed that her role included setting expectations for students to engage with the mathematics and creating a climate that nurtured these expectations.

How Stacey Enacted Her Curriculum Vision

Stacey described at least three aspects of how she planned with and enacted her curriculum vision:

1. She took advantage of the smaller unit booklets provided in one particular program.
2. She believed that to meet the district standards, she needed to insert skill work after the development of mathematical concepts.
3. She took seriously her role as facilitator to support students in high-level mathematical discussions.

One of Stacey's professional responsibilities was to order new curriculum materials for the middle school. She appreciated the flexibility offered by the smaller unit booklets in one particular program because the units focused on mathematical strands rather than sets of (often unrelated) topics. She consulted with the high school mathematics teachers to consider the vertical curriculum in this decision. With their recommendations and the district standards in mind, Stacey decided to purchase seventh-grade and eighth-grade units related to rational number and proportional reasoning, algebra, and area and volume. She used the teacher's guides, which provided information about how the units were related to one another and which suggested multiple ways in which key mathematical ideas in the units built on one another. She then considered how these units might fit together coherently.

Planning and Enacting Your Curriculum Vision

One important step in planning to enact a curriculum vision is to "look at your curriculum and begin to organize it into more coherent instruction" (Mirra 2008, p. 7). That is, examining the existing district or state curriculum guides in conjunction with available curriculum materials can provide a starting point for planning with and enacting a curriculum vision. When looking at curriculum materials, teachers might identify the curriculum vision (either implicit or explicit) of its authors. They can then consider how their curriculum vision is aligned with or different from that of the curriculum materials and how their standards fit into this vision. To achieve coherence across the grade levels, teachers should also consider their students' prior mathematics and what they will need to know for subsequent courses. At that time, decisions can be made about using and adapting the materials. Finally, it is useful for teachers to consider how their curriculum vision will influence their lessons and their interactions with students during instruction. The following questions might be asked when planning with and enacting a curriculum vision:

- What is the implicit curriculum vision of the authors of my curriculum materials?
- In what ways is my curriculum vision aligned with, or different from, that of my curriculum materials?
- How do the CCSSM and/or my state standards and district objectives fit into this vision?
- How do I take into account not only the horizontal curriculum (within my course) but also the vertical curriculum (across the grade band)?
- How can I use my curriculum materials as a tool to achieve my curriculum vision?
- Do I need to adapt, reorganize, or delete elements of the curriculum materials? If so, how?
- Do I need to supplement the curriculum materials with additional ideas or activities? If so, how?
- How does my curriculum vision impact the ways that I interact with my students?
- How do my individual students and their mathematical thinking influence my curriculum vision?

Although the new materials offered many high-level tasks organized toward building big mathematical ideas, Stacey was concerned about meeting the district's goals related to procedural fluency. Her experience taught her that it was easier to supplement high-quality, understanding-based curriculum materials with skill work than to build a rich, high-quality curriculum around a conventional, skill-based textbook.

Stacey's principal asked her how she figured out how to supplement the skill work. She carefully identified the topics in her local curriculum that were not covered in the materials. Then she determined where to insert direct instruction to cover these topics, keeping in mind the balance of developing big mathematical ideas along with promoting procedural fluency. She considered these questions:

- "Where is the best place to put that in?"
- "How much [skill work] is the best to put in?"

Describing the situation as hit and miss and inconsistent, she eventually found an appropriate balance through trial and error. Stacey's goal was to spend about two-thirds of class time working through the investigations and about one-third having students practice skills across a two-week time frame. She created practice sheets to follow each key concept and facilitated discussions that connected back to the big ideas of the investigations.

Stacey said that when first using the new curriculum materials, she was unsure about her approach. As she became comfortable with the high-level tasks and familiar with how students would respond to them, however, her vision related to her role and the role of her students became clearer. As her curriculum vision was refined, she began to concentrate on the "summarize" (Lappan et al. 1997) part of the lesson that follows the explorations in small groups. Typically carried out through whole-class discussion, this summary time highlights and pulls together key mathematical ideas. As Stacey worked at being explicit regarding what she expected from her students, their explanations and justifications became more detailed and conceptual over time. She found that having students talk about the mathematics helped her assess their growing understanding of the mathematical content. Ongoing formative assessments also helped Stacey pace her lessons and guide her planning for the next day. She found that continual adjustments were necessary to meet the needs of her individual students and support their mathematical understanding.

Conclusion

After reading the NCTM *Standards* (1989, 2000) and attending conferences, Stacey began to see that there was a different way to teach mathematics that was unlike her experiences as a student and a beginning teacher. She also used her experiences with a rich curriculum program and her knowledge of her students and mathematics to develop a curriculum vision that allowed her to enact coherent mathematics courses. Stacey continues to reflect on and develop her teaching, working to align her curriculum vision with her practice.

Many frequently changing factors influence classroom teachers' practices: national, state, and district standards; textbook adoptions; technological advancements; and research about student learning. One constant, however, is teachers' daily interactions with their students. Ultimately, the teacher must do what is best for his or her students. Taking Stacey's lead, teachers can use curriculum materials resourcefully and adapt them to their students in their own particular contexts. In fact, we believe that teachers are the only ones who can do this, since they are most familiar with their students, their school, and their community. Teachers need not do this work alone, however; developing curriculum vision can be a collaborative journey within and across mathematics departments.

References

Confrey, Jere, Alan P. Maloney, and Kenny H. Nguyen. *Learning Trajectory Display of the Common Core State Standards for Mathematics, Grades 6–8*. New York: Wireless Generation, 2010.

Darling-Hammond, Linda, James A. Banks, Karen Zumwalt, Louis Gomez, Miriam Gamoran Sherin, Jacqueline Griesdorn, and Lou-Ellen Finn. "Developing a Curricular Vision for Teaching." In *Preparing Teachers for a Changing World: What Teachers Should Learn and Be Able to Do*, edited by Linda Darling-Hammond and John Bransford, pp. 169–200. San Francisco: Jossey-Bass, 2005.

The El Barrio–Hunter College PDS Partnership Writing Collective. "On the Unique Relationship between Teacher Research and Commercial Mathematics Curriculum Development." In *Mathematics Teachers at Work: Connecting Curriculum Materials and Classroom Instruction*, edited by Janine T. Remillard, Beth A. Herbel-Eisenmann, and Gwendolyn M. Lloyd, pp. 118–33. New York: Routledge, 2009.

Grouws, Douglas A., Margaret S. Smith, and Paola Sztajn. "The Preparation and Teaching Practices of U.S. Mathematics Teachers: Grades 4 and 8." In *Results and Interpretations of the 1990 through 2000 Mathematics Assessments of the National Assessment of Educational Progress*, edited by Peter Kloosterman and Frank K. Lester, Jr. Reston, Va.: National Council of Teachers of Mathematics, 2004.

Lappan, Glenda, James T. Fey, William M. Fitzgerald, Susan N. Friel, and Elizabeth D. Phillips. *Getting to Know CMP*. Menlo Park, Calif.: Dale Seymour Publications, 1997.

Mirra, Amy. *Focus in Grades 3–5: Teaching with Curriculum Focal Points*. Reston, Va.: National Council of Teachers of Mathematics, 2008.

National Council of Teachers of Mathematics (NCTM). *Curriculum and Evaluation Standards for School Mathematics*. Reston, Va.: NCTM, 1989.

———. *Principles and Standards for School Mathematics*. Reston, Va.: NCTM, 2000.

National Mathematics Advisory Panel. *Foundations for Success: The Final Report of the National Mathematics Advisory Panel*. Washington, D.C.: U.S. Department of Education, Education Publications Center, 2008.

Chapter 7

Adapting Curriculum to Focus on Authentic Mathematics

Michelle Cirillo
Corey Drake
Beth Herbel-Eisenmann

TEACHERS' relationships with their curriculum materials often resembles, at best, "an arranged marriage" (El Barrio–Hunter College PDS Partnership Writing Collective 2009). In some cases, this arrangement is compatible and teachers feel comfortable using the materials as written; in many cases, however, it is not. At the same time, in an era of rigorous assessment and accountability procedures, teachers often face long lists of mathematics topics or learning expectations that need to be addressed across the school year (NCTM 2006). More specifically, classroom teachers are expected to address national standards "national standards (e.g., NCTM, 2000), state standards, and district standards…" by using mass-produced curriculum materials that were not written for their individual situations or students.

Despite these potential incompatibilities, mathematics teachers tend to rely heavily on commercial curriculum materials as their primary tool for teaching mathematics (Grouws, Smith, and Sztajn 2004). How do we learn to use and adapt impersonal curriculum materials in ways that allow us to meet our students' needs and address the multiple standards and objectives for which we and our students are responsible? What issues and questions might we consider in making decisions about our use of curriculum materials? In this chapter, we describe *curriculum vision*, which includes an understanding of the mathematics that students must learn and a coherent trajectory (or set of trajectories) for learning that mathematics. Teachers who manage the complex task of attending to students' needs as well as to standards and objectives should have a well-defined curriculum vision, and we offer suggestions for how they might begin to reflect on and articulate this vision.

Adapted from Cirillo, Michelle, Beth Herbel-Eisenmann, and Corey Drake. "Adapting Curriculum to Focus on Authentic Mathematics." *Mathematics Teacher* 103 (August 2009): 70–75.

This research was supported, in part, by the National Science Foundation under grant no. 0347906 (Beth Herbel-Eisenmann, principal investigator [PI]) and grant no. 0643497 (Corey Drake, PI). Any opinions, findings, conclusions, or recommendations expressed in this material are those of the authors and do not necessarily reflect the views of the National Science Foundation. We are also grateful to Matt for opening his classroom to us during this study.

Keys to Successful Teaching

Developing a curriculum vision is essential for teachers to be successful, researchers have suggested. For example, Darling-Hammond and others (2005, p. 177) claimed that "well-prepared teachers have developed a sense of 'where they are going' and how they and their students are going to get there." A teacher with a well-defined curriculum vision must have a deep understanding of (1) the curriculum's big mathematical ideas or goals, (2) the multiple trajectories students might follow to reach those goals, and (3) the strategies for using, adapting, supplementing, or creating curriculum materials to support students in reaching those goals. Finally, an important component of curriculum vision is curriculum coherence—mathematical connections that can be made by both students and teachers—within and across lessons.

Teachers should base their own curriculum vision on deep reflection on their knowledge, beliefs, and previous experiences in learning and teaching mathematics. In doing so, they should consider standards and objectives as well as the needs of particular students. Having an explicit curriculum vision can help teachers make strategic decisions about using and adapting curriculum materials flexibly and thus can help them achieve coherence.

A coherent mathematics curriculum is described in NCTM's Curriculum Principle: "A curriculum is more than a collection of activities: it must be coherent, focused on important mathematics, and well articulated across the grades" (NCTM 2000, p. 14). To promote students' deep understanding of mathematics, mathematical ideas must be linked to and build on one another (NCTM 2000). Because teachers tend to rely heavily on textbooks, however, they must, when working toward coherence, consider the curriculum vision put forth, either explicitly or implicitly, by the authors of their curriculum materials. As the National Mathematics Advisory Panel (2008) reported, the excessive length of U.S. mathematics textbooks (which often run 700 to 1000 pages) can contribute to a lack of coherence. By implication, teachers who follow the textbook too closely without making changes according to their own curriculum vision may be teaching a course that is not coherent.

Developing a Vision

The coherence that NCTM recommends and the Common Core State Standards for Mathematics (CCSSM) echoes can be achieved through the development and enactment of curriculum vision. First, it is important to examine each set of standards carefully, because the different perspectives that shape local, state, and national standards can result in either complementary or conflicting objectives across documents (Schmidt, McKnight, and Raizen 1997). When teachers consider the NCTM Standards and their relationship to other sets of standards, it is important to note that the Standards were not prepared as a "recipe for teachers or schools to follow" (Romberg 1998, p. 19). Rather, they were intended to provide a view of school mathematics curriculum that could be used to begin a dialogue with teachers, parents, administrators, and others to develop a plan for change (Romberg 1998). Thus, teachers should consider the ways in which the various standards documents fit into their curriculum vision within their own particular context.

Recognizing that different mathematical content ought to receive different emphases across grades 9–12 can help teachers think about the big ideas that students might learn throughout the year. For example, teachers might ask themselves, "How much emphasis should I place on functions in each of my courses, and what is a possible learning trajectory?" (cf. Confrey, Maloney, and Nguyen 2010). Another question relates to whether the curriculum is coherent and focused: "How can I create coherence or connections across these ideas, rather than treating them as a collection of topics? For example, how can I connect functions to regression analysis?" In addition, it is important to reflect on *how* the content will be taught: "What are the ways through which I want my students to acquire and use this content knowledge? For example, how might I teach theorems and properties of triangles in geometry and provide my students with opportunities to engage in the Mathematical Practices of CCSSM?"

In developing a curriculum vision, teachers might also consider their own beliefs about mathematics and their past experiences: "What do I see as the balance among skills, procedures, and concepts? How do my own knowledge, beliefs, and experiences fit into this vision?"

Since the Standards were not intended to be a "recipe," teachers must consider their own particular contexts as they develop curriculum vision: "How do my individual students' needs influence my curriculum vision? How do I consider particular features of high-quality mathematics education (e.g., NCTM's Principles for equity, technology, and so on) as guides and tools for making decisions about curriculum?" Many of these questions are summarized in figure 7.1.

Developing Curriculum Vision and Coherence

- What are the "big ideas" (the important mathematical content) that I want students to learn?
- Is my curriculum one that is coherent and focused rather than a collection of activities or topics? How do I create coherence or connections across these ideas?
- What are the ways through which I want my students to acquire and use this content knowledge? In other words, how can I teach this content while still addressing the important mathematical processes?
- What do I see as the balance among skills, procedures, and concepts?
- How do my own knowledge, beliefs, and experiences fit into this vision?
- How do my individual students' needs influence my curriculum vision?
- How do I consider particular features of high-quality mathematics education (e.g., NCTM's principles) as guides and tools for making decisions about curriculum?

Fig. 7.1. Questions to consider in developing a curriculum vision and coherence

In the next section, we use data from interviews to illustrate how one mathematics teacher, Matt, explored these questions as he developed his curriculum vision for a high school geometry course.

Case Study, Part 1: Matt Develops a Curriculum Vision

In his first year of teaching, Matt taught seventh- and eighth-grade mathematics, using a rich, Standards-based middle school curriculum program. Even after he changed districts and began teaching high school geometry, Matt's experiences with these materials continued to influence his developing curriculum vision:

> [The curriculum program] was very interesting to use. Basically [it] is like the model for me of what any textbook series should look like.... I blatantly stole a ton of stuff ... and copied it and gave it to [my geometry students] 'cause, I'm like, they're not gonna understand this unless they have some kind of basis for it.

When asked about that particular curriculum program being the ideal "model," Matt explained that students should be engaged with mathematics by answering questions and summarizing information rather than being *told* about mathematics.

Matt believed that the sequence of topics in the Standards-based program was "very well-constructed" and flowed together nicely. Because the program was coherent, his students did not even realize when they moved into a new topic. In contrast, using a more conventional textbook to teach his tenth-grade geometry course was a challenge for Matt because the book's curriculum vision did not support his own: "There's no coherence [in the geometry textbook]; there's no explanation. It's just, here's all your formulas in the boxes, now use them...." Formulas and theorems were isolated in green text boxes as finished results not necessarily connected to one other. This "presentation style" was not aligned with Matt's curriculum vision that students should be given opportunities to "make sense" of the mathematics.

Regarding proof, Matt objected to the way that the formulas and theorems in the textbook were simply "hand-waved at." Rather than communicating mathematics in a way that explained *why* formulas and theorems were true, the textbook, Matt believed, implied that students were simply supposed to accept the formulas and theorems as true. In his view, the textbook presented the material in this way: "Here it is. Accept it.... Let's ... not even discuss the reasons for it, let alone give the proof." Matt's main objection to this style was that it was not what he called "real math," which involved having students explore and conjecture, reason, prove, and explain their thinking, as advocated by NCTM's Standards.

Although Matt's curriculum vision was not aligned with the curriculum vision of the textbook that he was provided with, he was given a fair amount of freedom to decide how to teach geometry. In fact, the only directive that he was given by his school district was to "cover" certain chapters each semester. Matt's interpretation of this directive was that he could teach the content from these chapters in whatever way he thought best for his students.

In a later section, we explore data that provide insight into how Matt, a teacher attempting to enact a Standards-based curriculum vision of mathematics while using

Adapting Curriculum to Focus on Authentic Mathematics

conventional materials, implemented instruction that was both coherent and aligned with his vision. First, however, we introduce some questions designed to assist teachers in planning with and enacting their curriculum vision, because developing a coherent curriculum vision is only half of the curriculum vision equation.

Planning with and Enacting a Vision

One important step in planning to enact a curriculum vision is to "look at your curriculum and begin to organize it into more coherent instruction" (Mirra 2008, p. 7). A starting point is to examine the existing state or district curriculum guides in conjunction with available curriculum materials. Teachers might ask, "What is the curriculum vision implied by the authors of my curriculum materials?" Teachers can then consider how their curriculum vision is aligned with or different from that of the curriculum materials and how their state or district standards fit into this vision.

To achieve coherence across grade levels, teachers should also consider the mathematics that students have previously learned and the mathematics that they will need to know to be prepared for subsequent courses. In other words, they should ask themselves, "How does my curriculum vision fit into the horizontal curriculum [within my course] as well as the vertical curriculum [across grade bands]?"

Next, teachers will need to make decisions about using their curriculum materials to enact their curriculum vision: "How can I use my curriculum materials as a tool to achieve my curriculum vision? Do I need to adapt, reorganize, or delete elements of the curriculum materials? Do I need to supplement the curriculum materials with additional ideas or activities? If so, how?"

Finally, teachers will want to consider how their curriculum vision will influence enactment of their lessons: "How does my curriculum vision affect the ways that I interact with my students? In what ways will my students' thinking, contributions, and struggles influence my lessons?" We summarize these considerations for planning with and enacting a curriculum vision in figure 7.2.

Planning with and Enacting a Curriculum Vision

- What is the curriculum vision implied by the authors of my curriculum materials?
- In what ways is my curriculum vision aligned with or different from that of my curriculum materials?
- How do the standards and objectives of my state and district fit into this vision?
- How do I take into account not only the horizontal curriculum (within my course) but also the vertical curriculum (across the grade band)?
- How can I use my curriculum materials as a tool to achieve my curriculum vision?

Fig. 7.2. Questions to consider in planning with and enacting a curriculum vision

> - Do I need to adapt, reorganize, or delete elements of the curriculum materials? If so, how?
> - Do I need to supplement the curriculum materials with additional ideas or activities? If so, how?
> - How does my curriculum vision affect the ways in which I interact with my students?
> - How do my individual students and their communication about their mathematical thinking influence my curriculum vision?

Fig. 7.2.—*Continued*

In the next section, we return again to Matt and show how he planned with and enacted his curriculum vision.

Case Study, Part 2: Matt Plans with and Enacts His Vision

For Matt, the conventional textbook's curriculum vision initially acted as a "constraint" (Lloyd 2008, p. 66) on his teaching, because it was not aligned with his philosophy that mathematics makes sense and that students can and should understand mathematics. He felt the need to design activities that students could engage with in class in order to have the theorems "make actual sense to them instead of just being some random text in the green box."

Supplementing his textbook with additional activities allowed Matt to bring his practice closer to his professed beliefs about teaching proof. As his knowledge of the materials available to him developed over time, Matt planned with and enacted a curriculum vision that helped him achieve coherence. He created and adapted activities, adjusted the amount of time spent on various topics, and focused more explicitly on the process of proving rather than presenting theorems as finished results (see Cirillo [2008] for more detail).

Matt explained what he tried to do when enacting the written curriculum: "The written curriculum ... doesn't tell a story. It's like being given the *Cliffs Notes*, like here's the highlights.... What I try to do is ... make it seem cohesive and actually let it tell the story.... Let's see if we can figure out how these different pieces are related to each other."

Rather than presenting the material as a series of highlights, Matt's curriculum vision involved engaging students in activities and discussions that would assist them in understanding the mathematics and seeing the different connections that exist within mathematics.

Matt began to see the textbook as a resource that could be used to support the enactment of his own curriculum vision. In a lesson on the theorem that states that if two sides of a triangle are congruent, then the angles opposite those sides are congruent, Matt's textbook emphasized the theorem's algebraic applications rather than the proof process (see fig. 7.3). However, instead of simply telling students about the theorem and then having them solve simple exercises, Matt used the textbook's if-then statement of the theorem to engage his students in proving it. In many of his lessons, he would write a theorem on the board and ask students to draw a diagram for the proof. After agreeing on a diagram, students were then asked to use the diagram and the theorem's if-then format to write the "givens" and the "prove." Next, Matt asked the students to consider how they could prove the theorem. He gave them time to think about a proof as well as discuss their plan with a partner. Finally, in a whole-class discussion, students suggested plans for proving the theorems, and they wrote the proofs together.

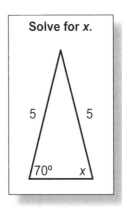

Fig. 7.3. An application of the base angles of an isosceles triangle theorem

Conclusion

Matt's previous experiences with a Standards-based curriculum program helped him see that there was "a different way"—different from the way that he had learned as a student of mathematics—to teach mathematics. He used his experiences with this program and his knowledge of his students and mathematics to develop a curriculum vision that allowed him to enact a more coherent geometry course, even when using a conventional textbook. Although Matt is not completely satisfied with how he is enacting his curriculum vision, he continues to reflect on and develop his teaching, working to align his curriculum vision with his practice.

Many factors that influence classroom teachers' practices—national, state, and district standards; new textbook adoptions; technological advancements; research about how students learn mathematics—change continually. One constant, however, is teachers' daily interactions with their students. Ultimately, the teacher's job is to do what is best for his or her students. Like Matt, through careful reflection on curriculum vision, teachers can use curriculum materials in resourceful and adaptive ways. In fact, we believe that teachers are the only ones who can do this kind of work, because they are the ones who know their students, their school, and their community. Teachers need not do this work alone, however. Developing curriculum vision can be a collaborative journey for an entire mathematics department.

The key is for everyone involved in mathematics teaching and learning to work toward the development of "curriculum capacity"—the combination of knowledge, beliefs, resources, vision, and support necessary to assist teachers in providing coherent mathematics instruction for all students (Drake 2009). And, while teachers' development of their own curriculum vision is one aspect of this capacity, we also call on curriculum designers to provide materials with more explicit curriculum vision, on district leaders to support teachers in developing vision and adapting materials, and on teacher educators to help teachers develop curriculum vision before entering the classroom.

References

Cirillo, Michelle. "On Becoming a Geometry Teacher: A Longitudinal Case Study of One Teacher Learning to Teach Proof." PhD diss., Iowa State University, 2008.

Confrey, Jere, Alan P. Maloney, and Kenny H. Nguyen. *Learning Trajectory Display of the Common Core State Standards for Mathematics, High School*. New York: Wireless Generation, 2010.

Darling-Hammond, Linda, James A. Banks, Karen Zumwalt, Louis Gomez, Miriam Gamoran Sherin, Jacqueline Griesdorn, and Lou-Ellen Finn. "Developing a Curricular Vision for Teaching." In *Preparing Teachers for a Changing World: What Teachers Should Learn and Be Able to Do*, edited by Linda Darling-Hammond and John Bransford, pp. 169–200. San Francisco: Jossey-Bass, 2005.

Drake, Corey. "Curriculum Capacity: Understanding District Investments in the Teacher-Curriculum Interaction." Paper presented at the Annual Meeting of the American Educational Research Association, San Diego, Calif., April 2009.

El Barrio–Hunter College PDS Partnership Writing Collective. "On the Unique Relationship between Teacher Research and Commercial Mathematics Curriculum Development." In *Mathematics Teachers at Work: Connecting Curriculum Materials and Classroom Instruction*, edited by Janine T. Remillard, Beth Herbel-Eisenmann, and Gwendolyn M. Lloyd, pp. 118–33. Studies in Mathematical Thinking and Learning, Alan H. Schoenfeld, series editor. New York: Routledge, 2009.

Grouws, Douglas A., Margaret S. Smith, and Paola Sztajn. "The Preparation and Teaching Practices of United States Mathematics Teachers: Grades 4 and 8." In *Results and Interpretations of the 1990–2000 Mathematics Assessments of the National Assessment of Educational Progress*, edited by Peter Kloosterman and Frank K. Lester, Jr., pp. 221–67. Reston, Va.: National Council of Teachers of Mathematics, 2004.

Lloyd, Gwendolyn M. "Curriculum Use While Learning to Teach: One Student Teacher's Appropriation of Mathematics Curriculum Materials." *Journal for Research in Mathematics Education* 39, no. 1 (January 2008): 63–94.

Mirra, Amy. *Focus in Grades 3–5: Teaching with Curriculum Focal Points*. Reston, Va.: National Council of Teachers of Mathematics, 2008.

National Council of Teachers of Mathematics (NCTM). *Principles and Standards for School Mathematics*. Reston, Va.: NCTM, 2000.

———. *Curriculum Focal Points for Prekindergarten through Grade 8 Mathematics: A Quest for Coherence*. Reston, Va.: NCTM, 2006.

National Mathematics Advisory Panel. *Foundations for Success: The Final Report of the National Mathematics Advisory Panel*. Washington, D.C.: U.S. Department of Education, Education Publications Center, 2008.

Romberg, Thomas A. "Comments: NCTM's Curriculum and Evaluation Standards." *Teachers College Record* 100, no. 1 (Fall 1998): 8–21.

Schmidt, William H., Curtis McKnight, and Senta A. Raizen. *A Splintered Vision: An Investigation of U.S. Science and Mathematics Education*. Dordrecht, Netherlands: Kluwer, 1997.

Section IV
Cultivating Mathematical Practices and Habits of Mind

Introduction

Although it is necessary to infuse courses and curricula with modern content, what is even more important is to give students the tools they will need in order to use, understand, and even make mathematics that does not yet exist. A curriculum organized around habits of mind tries to close the gap between what the users and makers of mathematics do and what they say. (Cuoco, Goldenberg, and Mark 1996, p. 376)

This quote from the authors' classic article, "Habits of Mind: An Organizing Principle for Mathematics Curricula," has roots in the NCTM's more general process standards and provided a conceptual foundation for the Standards for Mathematical Practice in the Common Core State Standards for Mathematics (CCSSM). The CCSSM's mathematical practices (fig. IV.1) include abilities, processes, and dispositions that enable students to understand mathematics and use mathematics with understanding.

Mathematical Practices

1. Make sense of problems and persevere in solving them.
2. Reason abstractly and quantitatively.
3. Construct viable arguments and critique the reasoning of others.
4. Model with mathematics.
5. Use appropriate tools strategically.
6. Attend to precision.
7. Look for and make use of structure.
8. Look for and express regularity in repeated reasoning.

Fig. IV.1. CCSSM Standards for Mathematical Practice

The three chapters in Section IV are written by the Education Development Center curriculum development team of Cuoco, Goldenberg, Mark, and Sword. The first three team members pioneered work using mathematical habits of mind that are central to the work of mathematicians and useful for organizing school mathematics curricula.

Regardless of the level at which you teach, you will find that each chapter's important message is relevant across the grades in this era of the CCSSM. The first chapter considers ideas of algebra, logic, techniques, and habits of mind as well as when and to what extent they can be learned with intellectual integrity in the elementary school grades before a formal study of algebra. In the second chapter, the authors argue that developing mathematical habits of mind in the middle grades is essential for making the critical transition from arithmetic to algebra. The authors of the third chapter reflect on their work in using the habits-of-mind approach for organizing a high school curriculum. They indicate that such an approach offers a vehicle for paring down the collection of methods and techniques one needs in high school, leaving a small set of general-purpose tools that tie together many seemingly different mathematical terrains.

Introduction

Building coherence in the development of mathematical ideas across the grades is a key to improving students' mathematical learning in the United States. Knowing the mathematical experiences, understanding, skills, habits of mind, and mathematical practices that students bring to a grade and what the expectations are for the following grades can help teachers bridge the transitions for students on each end.

As you read the three chapters in this section, think about their implications for better understanding the CCSSM's mathematical practices and what implementing some of those practices might look like in your classroom.

Questions for Reflection and Collective Discussion

1. Scan the *Common Core State Standards for Mathematics* (CCSSM) for grades K–5, noting instances of expected mathematical behaviors that involve ideas of early algebraic thinking. In the CCSSM, at what grade does the treatment of ideas of early algebra first occur? How does this compare with your current state mathematics standards? With current practices in the elementary school(s) in your district? Discuss your observations and possible instructional implications with your colleagues.

2. The first chapter emphasizes the role of children's intuition, logic, and linguistic knowledge in early learning about number, number operations, and early algebraic reasoning. As an elementary school teacher, how might you use the children's ways of thinking exemplified in this chapter in planning classroom implementation of the CCSSM number and algebra content standards for your grade?

3. Work with colleagues to create a chart summarizing how algebraic thinking is treated in the CCSSM grades 3–8 standards.

4. The second chapter discusses two algebraic habits of mind: *abstracting regularity from calculations* and *articulating a generalization using mathematical language*. These particular habits of mind, while very useful in algebra, are also applicable in the areas of number, geometry, statistics, and probability. Based on a mathematics class you are now teaching, give illustrations of how each habit of mind is applicable to the learning of a particular topic. Compare your illustrations with colleagues.

5. The authors of the third chapter discuss six general mathematical habits of mind that are an integral part of doing mathematics and should be a focus of teaching and learning school mathematics. They also discuss four habits of mind more prevalent in geometry and analysis. Seven additional habits of mind that are particularly appropriate in teaching and learning algebra are also described. Which habits of mind are cultivated in the mathematics classes you now teach? Share and elaborate on the habits of mind you identified. Which other habits of mind identified by the authors seem likely candidates for focus in the mathematics you teach?

6. In the second chapter, the authors assert that "much more important than specific mathematical results are the habits of mind used by the people who create those results." After reading and reflecting on the three chapters in this section, to what extent do you agree with the authors' assertion? Share your thinking with your colleagues.

7. How are the mathematical habits of mind illustrated in the three chapters of this section similar to, and different from, the process standards in your current state mathematics standards document?

8. Examine the descriptions of the CCSSM's mathematical practices (Common Core State Standards Initiative 2010, pp. 6–8). How are the mathematical practices similar to, and different from, the habits of mind discussed in the three chapters?

9. Look back at the chapter(s) you read in this section and identify specific mathematical practices that they exemplify. Compare your findings with those of your colleagues.

10. On the basis of your examination of the CCSSM's mathematical practices, identify practices you currently emphasize in your instruction and those you have not typically emphasized. Compare your lists with those of other teachers at your grade or of the same course. Identify places where professional development would be helpful.

11. What mathematical practices and mathematical habits of mind seem most appropriate for cultivation in the grade(s) or course(s) you teach? Share your choices and reasons with colleagues.

12. How could your school or district reshape its curriculum and instruction so that the mathematical practices become commonplace in every classroom?

References

Common Core State Standards Initiative. *Common Core State Standards for Mathematics*. Washington, D.C.: National Governors Association Center for Best Practices and the Council of Chief State School Officers, 2010. http://www.corestandards.org.

Cuoco, Al, E. Paul Goldenberg, and June Mark. "Habits of Mind: An Organizing Principle for Mathematics Curricula." *Journal of Mathematical Behavior* 15 (December 1996): 375–402.

Chapter 8

An Algebraic-Habits-of-Mind Perspective on Elementary School Mathematics

E. Paul Goldenberg
June Mark
Al Cuoco

Common wisdom tells teachers to introduce arithmetic first, algebra later. Reality is not so simple. Some algebraic ideas—for instance, those about the properties of binary operations apart from the numbers these operations may combine—develop naturally before children learn arithmetic. In fact, they must develop before arithmetic can make sense.

Using children's natural algebraic ideas to develop mathematical habits of mind can lead to deeper understanding in both algebra and arithmetic (Cuoco, Goldenberg, and Mark 1996; Cuoco, Goldenberg, and Mark 2010; Goldenberg 1996; Goldenberg, Shteingold, and Feuzig 2003; Goldenberg and Shteingold 2007; Mark et al. 2010). If children are to become competent at mathematics, including arithmetic, those habits of mind must take precedence over rules, formulas, and procedures that do not derive from logic that the child can grasp. The fact that algebraic ideas, logic, and techniques can be organized around the development of mind makes clear that we are truly talking about habits of mind rather than features of mathematics or idiosyncrasies of mathematicians. This chapter describes two of these natural habits of mind.

A Property of Addition before Addition

For young children, what will later be formalized as the commutative and associative laws of addition begins as an intuitive sense of stability, or invariance, of quantity under rearrangement. Piaget (1952) called it *conservation of number*; Wirtz and others (1962) and Sawyer (2003) called it the *any-order, any-grouping* property. Before conservation,

---

Adapted from Goldenberg, E. Paul, June Mark, and Al Cuoco. "An Algebraic-Habits-of-Mind Perspective on Elementary School." *Teaching Children Mathematics* 16 (May 2010): 548–56.

This research was supported, in part, by the National Science Foundation under grant nos. ESI-0099093, DRL-0733015, and DRL-0917958. The opinions expressed are those of the authors and not necessarily those of the NSF. Elaboration on the ideas in this chapter can be found at http://thinkmath.edc.org/index.php/EarlyAlgebra.

arrangement trumps number, but figure 8.1a may not have a fixed number associated with it. Later, the new conserver may not yet know how many fingers figure 8.1a shows without counting but will be sure that the number, whatever it is, stays put if the hands are moved as in figure 8.1b or even as in figure 8.1c. That algebraic idea, a property of aggregation, must exist before the arithmetic fact—knowing what number 2 + 5 is—can make sense.

Before children learn conservation of number, they may not associate a fixed number with an image.

(*a*) Later, new conservers may not yet know how many fingers are showing without counting them.

(*b*) But they will be sure that the number stays the same if the hands move this way.

(*c*) *And* even if the hands move as in this image, they will be sure that the number stays the same.

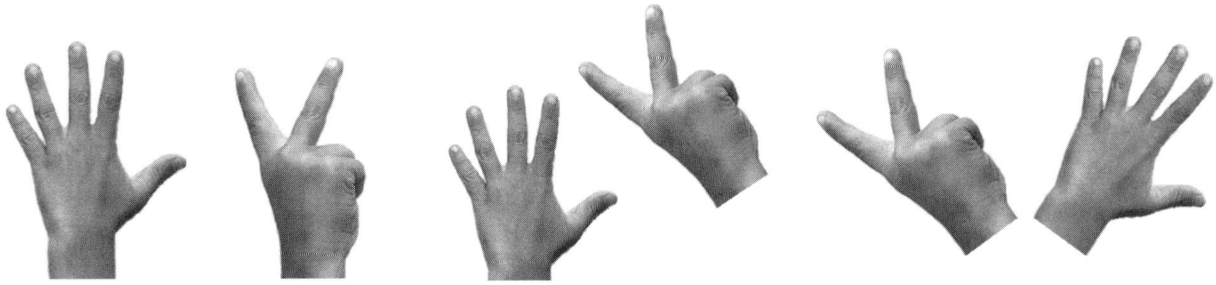

Fig. 8.1. Stability of a quantity under rearrangement

In a similar way, if an instructor hides a group of coins and asks, "How much money is there?" children for whom the question makes any sense will be absolutely certain that an answer exists and that only one answer is correct. They may be unsure about counting methods and may think that some methods might give incorrect answers, but conservers will know that just one correct answer exists. In fact, any child who really believes that the hidden amount can vary is not cognitively ready for the question of what the amount is. There is no "the amount" if it can vary. Most six-year-olds do not yet conserve number; by age seven, nearly all children do.

Having confidence that all three images in figure 8.1 represent the same quantity is not the same as knowing the commutative property, which is not about the arrangement of physical objects in space but about the behavior of a particular element (here, the +, or the plus sign) in a formal syntactic system of written symbols. In some contexts, children can make perfect sense of written symbols—even significant parts of algebraic notation—but most young children cannot make sense of formal operations on a string of symbols. So, at this early stage, commutativity remains largely an intuitively obvious idea about the physics of mathematics: the nature of aggregation, not the nature of symbols. Even so, educators can support a young child's logic better by recognizing that it already relies on the underlying ideas that formal mathematics will later codify. Children see that the principle applies regardless of the numbers. The principle captures the essential algebraic aspect of the structure of addition that commutativity is about.

Logical Precursors of the Distributive Property

Pick a number. Multiply it by five; also multiply your original number by two. Now add the results. You get the same answer you would get if you multiplied your original number by seven. A general statement of that fact, the distributive property, is possibly the most central idea in elementary school arithmetic, key to understanding the algorithms at the core of fluent mental calculations (e.g., 102 × 27 can be computed in two parts, as 100 × 27 + 2 × 27), and the logical basis for many rules of algebra that might otherwise seem arbitrary. This property relates multiplication and addition, but children know it long before they ever meet multiplication. The property is in the language (and logic) that youngsters use when they say that five (fingers, pennies, or 27s) plus two (fingers, pennies, or 27s) makes seven (fingers, pennies, or 27s). The following dialogues with six-year-olds late in their kindergarten year give a sense of what their logic does and does not handle. What distinguishes the questions the children get right from those they get wrong? What logic might explain the particular wrong answers they get? (**T** indicates the teacher's comments; **S1** and **S2** are both female students.)

> **T:** What's a really big number?
>
> **S1:** A million!
>
> **T:** Suppose I asked, "How much is a thousand plus a thousand?" What would you say?
>
> **S1:** [*with a big smile*] I have no idea!
>
> **T:** And suppose I asked, "How much is two thousand plus three thousand?"
>
> **S1:** [*thinking, then with confidence*] Five thousand!
>
> **T:** Suppose I asked, "How much is a hundred plus a hundred?" What would you say?
>
> **S2:** A hundred.
>
> **T:** What about, "How much is two hundred plus three hundred?"
>
> **S2:** Five hundred.
>
> **T:** [*playfully*] And what if I asked, "How much is a thousand plus a thousand?"
>
> **S2:** A million!

As soon as children are comfortable enough with the idea, the language, and the knowledge—perhaps late in kindergarten or early in first grade—to answer the question,

How much is three sheep plus two sheep? they will happily apply the idea, the language, and the knowledge to give the correct answer to the spoken question, How much is three hundred plus two hundred? or even to the question, How much is three-eighths plus two-eighths? However, what they have in mind may well be quite different from what adults have in mind when we give the same answer.

When teachers ask a slightly different question, How much is a hundred plus a hundred? (with no audible preceding small numbers such as two or three), young six-year-olds may repeat the words *a hundred* or say something such as *a million*.

If, instead, a teacher asks, "How much is an eighth plus an eighth?" little ones may give just a puzzled stare and no answer at all. If their arithmetic is strong enough, they might possibly count and answer *sixteen* (or, sometimes, *nine*).

Why such errors are made and why *hundred* and *eighth* lead to different errors are beyond the scope of this chapter. The point is that when no audible small numbers are given, little children tend to give wrong answers. But when the numbers are not too large, even some kindergartners tend to answer correctly; more first graders do; and we can absolutely count on it in second grade. Whatever an *eighth* or a *hundred* is, children are sure that three of them plus two of them is five of them! Although this does not constitute knowing the distributive property, it does tell us that the children already grasp the underlying idea that the distributive property will later formally encode.

If we use *sevens* (a fully understood fixed quantity) in place of *hundreds* (which may still be a nonspecific *zillions* for young children), youngsters still know that three of them plus two of them yields five of them. Once a child has a meaning for three sevens and that meaning is a specific number (even if the child does not yet remember which number), the child's long-standing logic, intuition, or linguistic knowledge that three sevens plus two sevens is five sevens becomes arithmetically usable.

The meaning might be given as an image (see fig. 8.2a), a sum (7 + 7 + 7), a product (3 × 7), or in other ways. Each way shows something *threeish* and something *sevenish*. Because 7 + 7 + 7 and 3 × 7 are both language, such expressions are best introduced as (mathematical) descriptions of a situation—for example, the array image—that communicate partly without analyzing the language formally. The image, of course, requires some analysis, too—visual rather than linguistic—to see the three sevens. To connect three sevens with twenty-one, we must agree that what makes figure 8.2b *seven* is its seven squares. Figure 8.2a is twenty-one because of its twenty-one squares, but it is also a picture of three sevens: a multiplication fact. Then, figure 8.2c shows that three sevens and two sevens makes five sevens.

The spoken form is familiar: "Three sevens plus two sevens makes five sevens." The pictures support the semantics of the situation, helping to establish the role of sevens and preserve its numerical meaning rather than letting it degenerate into a nonnumeric object, like sheep. In contrast, the classical written form—(3 × 7) + (2 × 7) = 5 × 7—is quite another story.

Such expressions as 7 + 7 + 7 and 3 × 7 are both language and are best introduced as mathematical descriptions that communicate partly without analyzing the language formally.

(a) Images require visual rather than linguistic analysis.
This array shows something *threeish* and something *sevenish*.
It is twenty-one because of its twenty-one squares, but it is also a picture of three sevens: a multiplication fact.

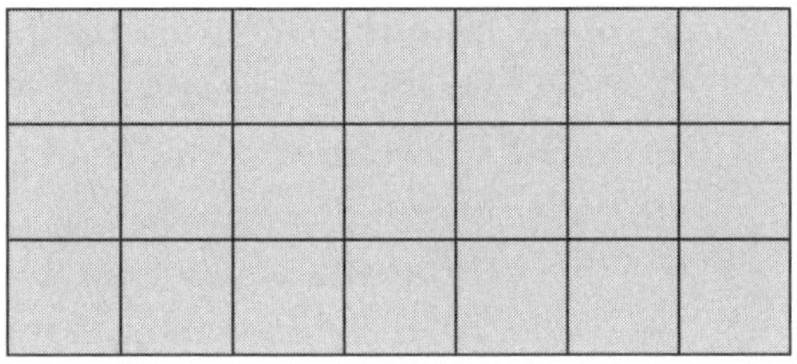

(b) To connect three sevens with twenty-one, we must agree that what makes this figure *seven* is its seven squares.

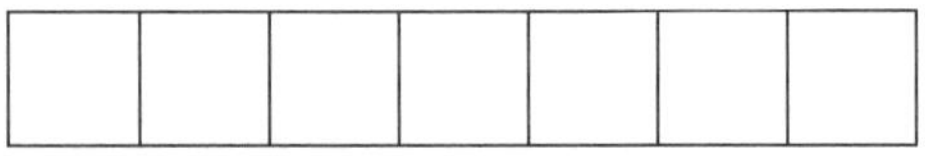

(c) Three sevens and two sevens makes five sevens.

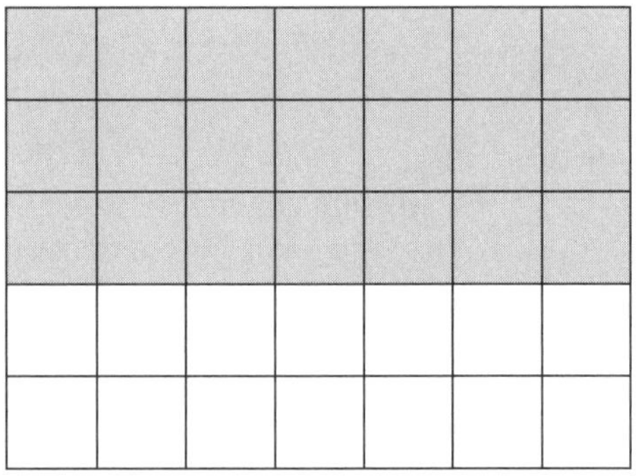

Fig. 8.2. Connecting visual representations with language descriptions

Spoken versus Written Symbols

A child's knowledge that the finger collections in figure 8.1 can be described by the same number does not guarantee that he or she will know that the print statements 5 + 2 and 2 + 5 refer to the same number. The written language of mathematics presents challenges that can be finessed by spoken language and by appropriate visual presentations. Perhaps the most glaring example is the canonically incorrect fourth-grade response to $3/8 + 2/8$ = ? Although $5/16$ is a common answer from fourth graders (and beyond), no first grader would ever respond, "Five-sixteenths." It is uninformative—in fact, misleading—to explain such errors simply by claiming that these expressions are too abstract or that children cannot handle symbols. Spoken words are symbols, too, and such words as *the*—which young children use flawlessly—are about as abstract as one can get. It is worth understanding the difference between figure 8.1 and 5 + 2 = 2 + 5 to see why the challenge of print for children may not be a mathematical challenge.

Humans have evolved to be quite flexible about visual order and orientation, but in the life of any individual human, it takes some learning. Infants who have come to recognize a bottle when it is handed to them in the proper orientation (see fig. 8.3a) do not, at first, reach for it when it is handed to them in some unfamiliar orientation, for example, with the nipple visible but facing away (see fig. 8.3b). Very soon they *do* learn to recognize objects regardless of their orientation. Considering the visual processing required, this is quite an impressive accomplishment. Even if the bottle is presented in the same orientation but at different distances, very different images are projected onto the retina. The distortion of parts relative to each other can be extreme, and yet babies recognize all these projections—most of which they have never seen before—as the same object.

Fig. 8.3. Recognizing objects regardless of orientation is a learned behavior.

Although this complex neural computation needs data (learning) to tune it up, the ability is hard-wired, an evolutionary gift essential for survival. Otherwise, we would have been meals for tigers that we did not recognize because they did not happen to be facing exactly the same way as when we first saw them. Our ancestors had to interpret different retinal images as being the same object as long as those images could be made *the same* under rotation, reflection, dilation, or certain projective transformations. As a result, our brains are adept at them.

An Algebraic-Habits-of-Mind Perspective on Elementary School Mathematics

But those ancestors did not read. The letters **d**, **b**, **q**, and **p** are all the same shape and differ only by rotation or reflection. To read, children must learn to see them as different objects, not as the same object in different orientations. So, young children's letter reversals are part of evolution's gift. To decode print, children must unlearn a principle that applies nearly everywhere else. They must treat print as an exception to the usual rules of seeing.

Moreover, **was** and **saw**—each just three print squiggles arranged in a different order—must *not* be recognized as the same. Alas, then come **2 + 5** and **5 + 2**, two perfectly good examples of print squiggles that *are* to be treated as the same. (As always, the truth is not so simple. On a number line, numbers represent addresses—the names of specific points or locations along the line—and also distances between addresses. The child who enacts **2 + 5**, perhaps by jumping along a large number line on the floor, would enact **5 + 2** differently.) It is therefore not surprising that the notation can cause confusion in some contexts, but this is an issue of notation, not of concept. Such written descriptions as (3 × 7) + (2 × 7) = (3 + 2) × 7 are typically opaque, unless they arise as abbreviations of language that the children themselves use to describe such displays as figure 8.2c.

The trouble is not with the underlying mathematical idea but with the notation through which it is communicated. In fact, the way instructors of kindergartners and early first graders teach writing can help here, too. Children tell stories. The teacher encodes their language in writing. For example, children say that

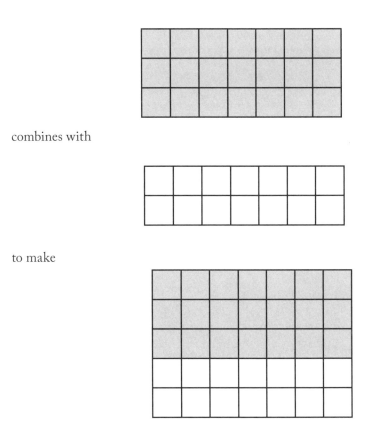

combines with

to make

As they speak, the teacher writes

$$(3 \times 7) + (2 \times 7) = (5 \times 7).$$

Getting arithmetically good enough to use this valuable property takes time and practice. But the underlying idea is part of the child's cognitive structure as soon as the child can meaningfully make such statements as, "Two sheep plus three sheep is five sheep." Again, the underlying idea must be there before any practice of it can make sense.

Possibly because of print's special status, the logic that children apply when information is presented in spoken symbols may not be applied when the same information is presented in print. The canonical error with fractions is a perfect example. The spoken question, "How much is three-eighths plus two-eighths?" focuses attention on three plus two and tends to evoke the correct reasoning and get the correct answer. By contrast, the written question does not focus attention on the top numbers only:

$$\frac{3}{8} + \frac{2}{8} = ?$$

Children for whom the meaning is not already established tend to interpret the plus sign as *add everything in sight*. Mathematical reading and writing are different from prose reading and writing. Prose flows strictly left to right, in one dimension. Bar and coordinate graphs, histograms, charts, tables, and so on are two-dimensional records. One must attend to horizontal and vertical positions to interpret them. Even symbolic expressions require attention to vertical position. That is obvious with fractions

$$\frac{3}{8} + \frac{2}{8}$$

but is also true of 3^2, which is not the same as 32. Even mathematical writing that is only horizontal cannot be read strictly left to right. Mathematical writing like the following expression and equation requires attention to the right side before attention to the left.

$$2 \times (3 + 5)$$

and

$$7 + \underline{} = 5 + 4$$

In fact, some expressions, like 7 + 6 ÷ 2, require both left-to-right and right-to-left analysis. In this example, 6 ÷ 2 must be evaluated left to right (because 2 ÷ 6 is different), and yet the convention about order of operations dictates that the whole right side of the expression, 6 ÷ 2, be evaluated before the addition that is specified by the 7 + part.

Algebra as a Language

Algebraic notation is used in two distinct ways: for describing what we know and for deriving what we do not know. In the former, algebra is a language for describing the structure of a computation, a numerical pattern we have observed, a relationship among varying quantities, and so on. Young children are phenomenal language learners.

Exercises such as the one in table 8.1 (but without the leftmost column) are familiar enough in many curricula. Children look for a pattern in the inputs and outputs, determine a rule, and complete the table. The Think Math! curriculum (EDC 2008) often adds a pattern indicator (the leftmost column) to problems of this kind.

Table 8.1
A Pattern Indicator Gains Meaning from Context When It Accompanies a Find-the-Rule Exercise.

n	10	8	28	18	17			58	57
$n-8$	2	0	20			3	4		

Michelle, a second grader in a classroom using Think Math!, completed table 8.1 before her teacher had finished distributing copies to all the children. When the teacher asked how she had done it so fast, Michelle answered, "Well, I saw it was take away eight, because I looked at the twenty-eight and the twenty, and then I saw that ten and two was take away eight again, and then I saw eight and zero." Pointing to the leftmost column and grinning as if the teacher had left a clue by mistake, Michelle exclaimed, "Besides, it *says* it right here!"

How did she know? Nobody had ever discussed *variables* or *letters standing for numbers* or had even mentioned that first column. If Michelle had seen *only* table 8.2, with no examples to infer from, she most likely would not have felt that the symbols say anything. But having discovered the pattern, she thought that the symbols looked close enough to mean the same thing, so she *then* assigned them that meaning.

Table 8.2
A Pattern Indicator Without a Pattern from Which to Infer Its Meaning Would Be Simply More to Learn.

n	18	17			58	57
$n-8$			3	4		

In other words, she did what little children excel at: She learned language (in this case, $n-8$) from context. If algebraic language is part of the environment, used where context gives it meaning, children can apply their natural—and extraordinary—language-learning prowess to it and learn to use it descriptively. Just as children learning their native language understand, at first, more than they can say, Michelle could not immediately produce such descriptive language, but she and others try these interesting ways of writing what they know and, over time, become good at it. For instance, fourth graders learn this trick:

Think of a number; add three; double that; subtract four; cut it in half; subtract your original number; your result is one.

They love the trick and want to show their parents and friends. They also want to know how it works. To explain, we can add pictures (see fig. 8.4). The act of doubling, which most fourth graders find quite natural and obvious, is, again, the distributive property in action. Although the expression $2(b + 3)$ does not make obvious what the result is, children readily learn to describe the third picture (see fig. 8.4c) as *two bags plus six* and abbreviate that description as $2b + 6$. They do not have to talk about variables or letters standing for numbers; they simply describe what they know and then write it as simply as they can. See a detailed description of this algebraic thinking with children on the Think Math! Web site (EDC 2009), and see Sawyer (1964) for the original source of the idea. Furthermore, Mark and her colleagues describe yet another way in which Think Math! gives students this algebra-as-description-of-what-you-know experience (Mark et al. 2012).

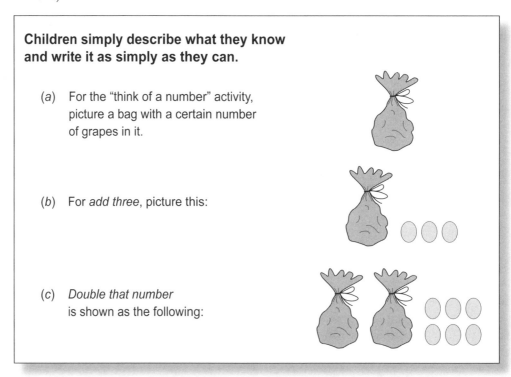

Fig. 8.4. Algebraic thinking without mathematical symbols

So Why Not Teach Algebra in Grade 4?

The other use of algebra—deriving what we do not know—requires formal syntactic operation on a set of symbols. Formal operations on strings of algebraic symbols—rearranging them, apart from their semantics, to create other strings of symbols that solve a problem—are, well, formal operations, and children are not, by and large, formally operational before age eleven and not reliably so before about age thirteen. It is not because fourth graders cannot handle symbols or abstract ideas—words are symbols; pictures are symbols; little children can be symbolic and abstract from babyhood—but because of the need to manipulate the symbols apart from their semantics. Children are generally unable to divorce symbols from meanings before roughly age twelve, so algebra as a course is delayed until then.

However, only the part of algebra that requires deduction by formal rules must wait that long. The part of algebra that is expressive of what we already know—that is, essentially, shorthand for semantic content clearly tied to a context we already understand—*that* part can be learned earlier. It is just language to express oneself, and children are excellent language learners. They do not learn language from explanations or formal lessons; they learn it from use in context. And, if it is learned all along, as it becomes developmentally possible, then, when the child is in late middle school, the transition to the new use of that language for deductive purposes could, presumably, be much easier, much more accessible for all children, much less of a brick wall of a million seemingly new things to learn all at once.

What about Elementary School?

Taking advantage of children's natural algebraic ideas and honing them is a focus on habits of mind rather than on rules that can otherwise seem arbitrary. The precursors of commutative and distributive properties described earlier must be refined, honed, extended, practiced, codified, and generalized, but they are already there as natural logic, the child's natural habits of mind and the building blocks of higher mathematics. If children are to become competent at mathematics, including arithmetic, those habits of mind must take precedence over rules, formulas, and procedures that do not derive from logic that the child can grasp. In fact, children can grasp a lot more if the foundations for their learning are grounded in their logic, which gives students all the tools to understand, not just memorize, the algorithms for arithmetic with whole numbers and fractions. The dramatically disappointing result of learning rules apart from understanding is the tendency to easily get mixed up and use procedures that do not work (Carpenter et al. 1997).

Organizing the arithmetic part of the elementary school mathematics curriculum around mathematical habits of mind would not shift the curriculum dramatically in content, except to give more attention to mental arithmetic than is usual. Paper-and-pencil methods are engineered to make the work easy and to reduce the cognitive load of calculation, the amount of thinking one needs to do. Judiciously chosen mental arithmetic tasks both exercise and depend on mathematical ways of thinking that the paper-and-pencil algorithms deliberately try to avoid—mathematical ways of thinking that are the backbone of the successful preparation for algebra that we want for our students. What would shift if we were to emphasize habits of mind is the order in which students acquire content? Instead of being the preparatory step for computing, algorithms would become the culmination of understanding how the computation works, another case of describing what we already know and abbreviating that description. Taking full advantage of the natural logic and algebraic ideas of young learners, and helping them refine and communicate those ideas in mathematical language, would produce students who are better at arithmetic as well as better prepared for and familiar with algebra.

References

Carpenter, Thomas P., James Hiebert, Elizabeth Fennema, Karen C. Fuson, Diana Wearne, and Hanlie Murray. *Making Sense: Teaching and Learning Mathematics with Understanding.* Portsmouth, N.H.: Heinemann, 1997.

Cuoco, Al, E. Paul Goldenberg, and June Mark. "Habits of Mind: An Organizing Principle for Mathematics Curriculum." *Journal of Mathematical Behavior* 15, no. 4 (December 1996): 375–402.

———. "Organizing a Curriculum around Mathematical Habits of Mind." *Mathematics Teacher* 103, no. 9 (May 2010): 682–88.

Education Development Center (EDC). Think Math! Comprehensive K–5 Curriculum. Boston: School Specialty Math, 2008.

———. Think Math! "Algebraic Thinking." 2009. http://thinkmath.edc.org/index.php/Algebraic_thinking.

Feigenson, Lisa, Susan Carey, and Elizabeth Spelke. "Infants' Discrimination of Number vs. Continuous Extent." *Cognitive Psychology* 44, no. 1 (February 2002): 33–66.

Goldenberg, E. Paul. "'Habits of Mind' as an Organizer for the Curriculum." *Journal of Education* 178, no. 1 (1996): 13–34. Reprinted as "Hábitos de pensamento; um princípio organizador para o currículo(I)." *Educação e Matemática* 47 (March–April 1998): 31–36; and "Hábitos de pensamento; um princípio organizador para o currículo (II)." *Educação e Matemática* 48 (May–June 1998): 37–44.

Goldenberg, E. Paul, June Mark, and Al Cuoco. "The Algebra of Little Kids: Language, Mathematics, and Habits of Mind." 2009. http://thinkmath.edc.org/index.php/Early_algebra.

Goldenberg, E. Paul, Nina Shteingold, and Nannette Feurzig. "Mathematical Habits of Mind." In *Teaching Mathematics through Problem Solving: Prekindergarten–Grade 6*, edited by Frank K. Lester, Jr., and Randall I. Charles. Reston, Va.: National Council of Teachers of Mathematics, 2003.

———. "The Case of Think Math!" In *Perspectives on the Design and Development of School Mathematics Curricula*, edited by Christian R. Hirsch, pp. 49–64. Reston, Va.: National Council of Teachers of Mathematics, 2007.

Gopnik, Alison, Andrew N. Meltzoff, and Patricia K. Kuhl. *The Scientist in the Crib: What Early Learning Tells Us about the Mind.* New York: HarperCollins, 2000.

Mark, June, Al Cuoco, E. Paul Goldenberg, and Sarah Sword. "Developing Mathematical Habits of Mind." In *Curriculum Issues in an Era of Common Core State Standards for Mathematics*, edited by Christian R. Hirsch, Glenda T. Lappan, and Barbara J. Reys, pp. 105–110. Reston, Va.: National Council of Teachers of Mathematics, 2012.

Piaget, Jean. *The Child's Conception of Number.* London: Routledge and Kegan Paul, 1952.

Sawyer, Warwick W. *Introducing Mathematics 1: Vision in Elementary Mathematics.* New York: Penguin Books, 1964. Reprinted, New York: Dover Publications, 2003.

Sfard, Anna. *Thinking as Communicating.* New York: Cambridge University Press, 2008.

Wirtz, Robert W., Morton Botel, M. Beberman, and Warwick W. Sawyer. *Maths Workshop: A Worktext for Children.* Toronto, Chicago: Encyclopedia Britannica Press, 1962.

Chapter 9

Developing Mathematical Habits of Mind

June Mark
Al Cuoco
E. Paul Goldenberg
Sarah Sword

> *Mathematical habits of mind* include reasoning by continuity, looking at extreme cases, performing thought experiments, and using abstraction that mathematicians use in their work (Cuoco, Goldenberg, and Mark 1996; Goldenberg 1996). Current recommendations emphasize the critical nature of developing these habits of mind: "Once this kind of thinking is established, students can apply it in the context of geometry, trigonometry, calculus, data and statistics, or other advanced courses." (Achieve 2008, p. 4)

THE RECENTLY published *Focus in High School Mathematics* (National Council of Teachers of Mathematics [NCTM] 2009) suggests that curricular approaches emphasizing sense making and reasoning are necessary foundations for the content taught in high school. We argue that developing mathematical habits of mind in the middle grades is essential for students who are making the critical transition from arithmetic to algebra. These habits can also build on students' early algebraic thinking and help them move toward a more formal investigation of algebraic ideas typically found in an eighth-grade or ninth-grade algebra course. For more than a decade, we and others have been engaged in curriculum development efforts that aim to provide students with the kinds of experiences that will help develop these mathematical habits and put them into practice (Cuoco 2007, 2008; EDC 2008, 2009; Goldenberg, Shteingold, and Feurzig 2007).

Much more important than specific mathematical results are the habits of mind used by the people who create those results. We envision curricula that elevate the methods by which mathematics is created and the techniques used by researchers to a status equal to that enjoyed by the results of that research. Our goal is not to train large numbers of students to be university mathematicians, but rather for students to become comfortable

---

Adapted from Mark, June, E. Paul Goldenberg, Al Cuoco, and Sarah Sword. "Developing Mathematical Habits of Mind." *Mathematics Teaching in the Middle School* 15 (May 2010): 505–9.

solving mathematical problems, to see the benefit of being systematic and using abstract reasoning, and to look for and develop new ways of describing situations. We want students to gain the tools they will need to use, understand, and even make mathematics that does not yet exist.

The Transition from Arithmetic to Algebra

Through our curriculum development work, we found overwhelming uniformity in mathematics teachers' assessments of the areas that cause students to struggle with algebra. Students appear to have trouble with the following aspects of mathematics:

- Expressing generality with algebraic notation, especially prevalent when students set up equations to solve word problems
- Understanding slope, graphing lines, and finding equations of lines
- Making sensible and fluent calculations with fractions
- Understanding and using proportional reasoning
- Building and using algebraic functions

At first read, this list looks like a collection of disparate topics. However, the common thread running through these topics lies in the mode of thinking required to master them. The key to such mastery are the two mathematical habits of—

1. abstracting regularity from calculations and
2. articulating a generalization using mathematical language

These skills are extremely effective in helping students transition from arithmetic to algebra. By "effective," we mean that students who succeed in algebra have developed these habits, whereas students who struggle at the cusp between arithmetic and algebra have not (Koedinger 2002).

Our experience has shown that as students become more practiced at repeated calculations, they gain the experiential foundation for identifying a regular pattern in their mental work. They also begin to see calculations, and the patterns they embody, as part of a larger algebraic process. In so doing, they are able to connect prealgebraic calculations to an algebraic process.

Two Key Algebraic Habits of Mind

Algebra as a scientific discipline evolved in an attempt to express general properties of binary operations (such as addition and multiplication) without having to refer to specific inputs to those operations. The need to abstract from calculations is essential to success in algebra. In the following example, the skill of abstracting from calculations is combined with the habit of expressing the abstraction with algebraic symbols to solve the problem. Using the *guess, check, and generalize* method makes a challenging topic (such as solving word problems) much easier for students.

Example 1: Comparing CD Prices

Suppose you want to buy a music CD. A Web site offers a 28 percent discount on the list price. It also adds 5 percent for state sales tax and $3.50 for shipping. The local music store sells CDs for 10 percent off the list price, also charges 5 percent sales tax, but has no shipping charge. Ignoring convenience and drive time, is it less expensive to buy online or to buy from the local music store?

Many ad hoc ways of solving such problems are possible, but a core algebraic approach is to develop a function or algorithm that will allow the student to easily compute the price of a CD in each scenario. For example, when buying from the Web site, the total cost of a CD is determined by taking the list price, subtracting 28 percent, then adding 5 percent and $3.50.

This may sound simple enough, but students who have difficulty with algebra are often unable to articulate this process without referring to a specific price, such as $25. In other words, they can compute the total cost of a CD for any *particular* list price, but cannot express the general process for calculating the total cost of a CD. That process is precisely the first step needed to find the break-even point (either by trial and error or by solving an equation).

One way to develop the skill of expressing generality is to begin with some numerical examples until a student "gets the rhythm" of the calculations and is able to articulate the cost of a CD in a way that will work for any price. If shopping on the Web, the thought development might proceed in this way.

If the CD lists for $25, this is my cost:

$$\$25 - (0.28 \cdot 25) = \$18, \text{ then}$$
$$1.05 \cdot \$18 = \$18.90, \text{ then}$$
$$\$18.90 + \$3.50 = \$22.40.$$

If the CD lists for $30, this would be my cost:

$$\$30 - 0.28 \cdot 30 = \$21.60, \text{ then}$$
$$1.05 \cdot \$21.60 = \$22.68, \text{ then}$$
$$\$22.68 + \$3.50 = \$26.18.$$

Eventually, after several more examples, you can start to "chain" the calculation; a $20 CD is used below:

$$1.05 \cdot [\$20 - (0.28 \cdot 20)] + \$3.50$$

With a few more examples, the general process emerges. The actual value of the list price does not matter; the steps are always the same. Using a placeholder c for the list price, the process continues:

or

$$1.05(c - 0.28 \cdot c) + \$3.50$$

$$1.05 \cdot 0.72c + \$3.50$$

We want to focus on the dual skills of (1) expressing this form of generality with algebraic symbolism and (2) transforming the resulting expressions to solve problems. In this example, the corresponding algorithm for the music store is

$$1.05 \cdot 0.9 \cdot c.$$

To determine the break-even point, find the value of c so that

$$(1.05 \cdot 0.72c) + \$3.50 = 1.05 \cdot 0.9c.$$

This method—called *guess, check, and generalize*—is not meant to be a guess-and-check method. The goal is for the student to realize that he or she is always repeating the same arithmetic calculations (with different numbers) and is then *generalizing* from those calculations. Because word problems are common in most middle school mathematics materials, students have myriad opportunities to practice the skill of guess, check, and generalize and hone their abilities to use these two key algebraic habits of mind.

Students can use this approach for any word problem they encounter. For example, there is no need for specialized approaches when problems involve rate or mixture. The real beauty is that it allows students to develop the powerful mathematical habits of mind of *abstracting regularity from calculations* and *articulating a generalization using mathematical language*.

Example 2: Multiplication Patterns

The next example shows how algebraic language is used to describe a discovery that students make. In the abstract, we would describe it as an *algebraic identity*. Students reexamine multiplication facts with a pattern-seeking mindset. They conduct elementary mathematical "research"—looking for a mathematical pattern and then extending and generalizing it—to practice their algebraic-descriptive skills and gain greater computational fluency. (See fig. 9.1 for an example of students' work using a number line.)

Students learn that notation is another way of describing what they already know, rather than it being an arbitrary code that means "you plug in the number where you see the letter." This learning process occurs in essentially the same way in which all language is learned.

This learning situation is a case of abstracting a *common structure* from a repeated calculation. Students make the abstraction naturally and *demonstrate* that they have already made the abstraction by answering 2499 when presented with 49 • 51. The new and harder step is to articulate what they already know: The algebraic "trick" of giving names to numbers helps make the description easier, first in English and then abbreviated in algebraic style.

This process allows students to transition from performing concrete calculations to noticing the structure of those calculations and describing that structure in algebraic language. Students get the clear message that mathematics has the power to *explain*, and that symbolism is a *help*, not just an extra task to learn. Students develop that algebraic language not from a set of arbitrary symbols and rules but as a natural way of expressing what they already know. In so doing, they find a simpler and less cumbersome way to express mathematical ideas than that provided by spoken language.

Students start by comparing two multiplication problems on a number line.

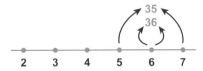

The teacher structures which pairs to compare, but the students name the numbers to enter. Students create several pairs.

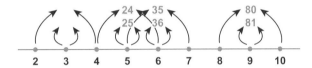

Most students notice a pattern: The numbers in each pair differ by 1. They are delighted when they discover that they can multiply 49 × 51 in their heads by *using* the pattern that they have abstracted from this experiment.

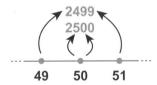

But *describing* this abstraction in English proves to be difficult. Students may remark, "If you take a number and multiply it by itself, then, if you take the next number and multiply it by the number just before the number...." But when we give *names* to the number—perhaps the name of the student—it gets much easier. "If you take a number, let's call it Dana, then (Dana − 1) × (Dana + 1) is one fewer than Dana × Dana." That can be abbreviated to

$$(d - 1) \times (d + 1) = (d \times d) - 1,$$

or simply

$$(d - 1)(d + 1) = d^2 - 1.$$

(For a full description of this activity, see http://thinkmath.edc.org/index.php/Difference_of_squares.)

Fig. 9.1. A number line is used to help students visualize comparisons and pairings.

Conclusion

We have used these examples to illustrate two particularly powerful mathematical habits of mind that focus on the critical transition from arithmetic to algebra. These habits take advantage of students' natural inclinations to find patterns and describe those patterns with language. The middle grades are an especially good time developmentally to build these skills. Students are certainly able to use language adeptly to express and describe mathematical ideas, which they must do before they can become comfortable with formal operations and working with the more formal syntactic and deductive aspects of algebra.

References

Achieve, Inc. "Policy Brief: The Building Blocks of Success: Higher Level Math for All Students." Washington, D.C.: Achieve, 2008. http://www.achieve.org/node/844.

Cuoco, Al. "The Case of the CME Project." In *Perspectives on the Design and Development of School Mathematics Curricula*, edited by Christian R. Hirsch, pp. 119–28. Reston, Va.: National Council of Teachers of Mathematics, 2007.

———. "Introducing Extensible Tools in Middle School and High School Algebra." In *Algebra and Algebraic Thinking in School Mathematics*, Seventieth Yearbook of the National Council of Teachers of Mathematics (NCTM), edited by Carole Greenes, pp. 51–62. Reston, Va.: NCTM, 2008.

Cuoco, Al, E. Paul Goldenberg, and June Mark. "Habits of Mind: An Organizing Principle for Mathematics Curriculum." *Journal of Mathematical Behavior* 15, no. 4 (December 1996): 375–402.

Education Development Center. *Think Math!* Orlando, Fla.: Houghton Mifflin Harcourt School Publishers, 2008.

———. *The CME Project*. Boston: Pearson Education, 2009.

Goldenberg, E. Paul. "'Habits of Mind' as an Organizer for the Curriculum." *Journal of Education* 178, no. 1 (1996): 13–34. Reprinted as "Hábitos de pensamento; um princípio organizador para o currículo (I)." *Educação e Matemática* 47 (March–April 1998): 31–36; and "Hábitos de pensamento; um princípio organizador para o currículo (II)." *Educação e Matemática* 48 (May–June 1998): 37–44.

Goldenberg, E. Paul, Nina Shteingold, and Nannette Feurzeig. "Mathematical Habits of Mind for Young Children." In *Teaching Mathematics through Problem Solving: Prekindergarten–Grade 6*, edited by Frank K. Lester, Jr., and Randall I. Charles, pp. 15–29. Reston, Va.: National Council of Teachers of Mathematics, 2003.

———. "The Case of Think Math!" In *Perspectives on the Design and Development of School Mathematics Curricula*, edited by Christian R. Hirsch, pp. 49–64. Reston, Va.: National Council of Teachers of Mathematics, 2007.

Koedinger, Kenneth R. "Toward Evidence for Instructional Design Principles: Examples from Cognitive Tutor Math 6." In *Proceedings of the Twenty-fourth Annual Meeting of the North American Chapter of the International Group for the Psychology of Mathematics Education*, Vol. 1, edited by Denise S. Mewborn, Paola Sztajn, Dorothy Y. White, Heide G. Wiegel, Robyn L. Bryant, and Kevin Nooney, pp. 21–49. Columbus, Ohio: ERIC Clearinghouse for Science, Mathematics, and Environmental Education, 2002.

National Council of Teachers of Mathematics (NCTM). *Focus in High School Mathematics: Reasoning and Sense Making*. Reston, Va.: NCTM, 2009.

Chapter 10

Organizing a Curriculum around Mathematical Habits of Mind

Al Cuoco
E. Paul Goldenberg
June Mark

IN THE YEARS immediately following the release of *Curriculum and Evaluation Standards for School Mathematics* (NCTM 1989), there was a strong imperative to invigorate school mathematics. With government and private funding, teams around the country took a clean-slate approach to precollege mathematics and developed new curricula aimed at making more mathematics more meaningful and more accessible to more students. A central focus of these efforts was a shift toward student-centered, problem-based classrooms in which teachers support students as they work out the mathematics for themselves.

At the high school level, along with this shift in pedagogy came topical reorganizations—the elimination of topics considered less useful than in previous generations and the infusion of more "modern" topics that had found or that had shown potential to find applications in a variety of mathematics-related fields.

Around 1992, the authors started thinking about alternatives to this approach to reform. It seemed to us that the real utility of mathematics for many students, especially for those who would not go into STEM fields, came from a style of work—a web of ways of thinking about the world—that mathematicians use in their profession. Mathematicians have long realized that their methods are often just as important as their results. For example, William Thurston (1994, p. 176) notes: "What mathematicians most wanted and needed from me was to learn my ways of thinking, and not in fact to learn my proof of the geometrization conjecture for Haken manifolds."

Our teaching experience and our work with teachers convinced us that raising the methods used by mathematicians to the same level of importance as the results of those methods would be a viable organizer for a high school curriculum. We made a detailed analysis of what we came to call *mathematical habits of mind*, and we made a case for

---

Adapted from Cuoco, Al, E. Paul Goldenberg, and June Mark. "Organizing a Curriculum around Mathematical Habits of Mind." *Mathematics Teacher* 103 (May 2010): 682–88.

This work was supported, in part, by the National Science Foundation under grant nos. ESI-0099093, DRL-0733015, and DRL-0917958. The opinions expressed are those of the authors and not necessarily those of the NSF.

using the ways of thinking that are indigenous to mathematics as a central benchmark both for deciding what topics to include in a high school curriculum and for determining how to develop them (Cuoco, Goldenberg, and Mark 1996).

Here we discuss the implications of the habits-of-mind approach for high school curricula, using the Center for Mathematics Education (CME) Project (EDC 2009) as a source of examples. In the course of developing this program, we found that using mathematical habits of mind as an organizer can bring genuine and often surprising coherence to a curriculum (Cuoco, Goldenberg, and Mark 2009). Although our original motivation was to help students make sense of the mathematical topics they study, the habits-of-mind approach also provided us with a vehicle for paring down the collection of methods and techniques one needs in high school, leaving a small set of general-purpose tools (Cuoco 2008) that tie together many seemingly different mathematical terrains.

Habits of Mind

We will examine general mathematical habits that are used across high school as well as specific geometric and algebraic habits. Good descriptions of statistical thinking and how it differs from mathematical thinking can be found in the Scheaffer and Tabor chapter in this volume, in *Focus in High School Mathematics: Reasoning and Sense Making* (NCTM 2009a) and in Cobb and Moore (1997).

General Mathematical Habits

These ways of thinking run throughout mathematics and can be developed within the contexts of algebra, geometry, probability, and statistics. They include the following.

Performing Thought Experiments

One can ask many questions about the classic context in which one starts with a rectangle of a certain size and cuts small congruent squares out of each corner, folding up the sides to make an open box. To reason about this context, one needs to be able to have a mental image, seeing, perhaps, a continuum of boxes starting out as a flat base with no height and morphing into a folded sheet of paper. Developing a knack for picturing such phenomena takes time and practice. Physical or computational manipulatives can help students develop this knack, but they are no substitute for the ability to create and perform thought experiments on one's own. One goal of our curriculum materials is to make explicit the kinds of thought experiments that bring life to such contexts.

Finding, Articulating, and Explaining Patterns

Finding regularity, especially subtle regularity, is a prized and useful skill in mathematics. So is articulating in precise language what one sees. And, of course, explaining why things happen is at the heart of doing mathematics.

For example, consider the portion of the multiplication table shown in figure 10.1, oriented in a way that agrees with the typical orientation of Cartesian coordinates (EDC 2009). Readers are invited to—

- find a recurring pattern in the table;
- describe that pattern in precise mathematical language; and
- explain why that pattern exists.

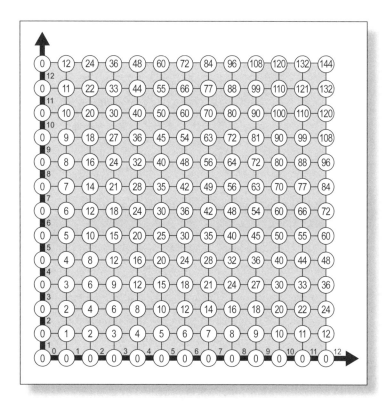

Fig. 10.1. Finding, describing, and explaining patterns is an important habit of mind.

Creating and Using Representations

Much has been written about the problem of counting the number of "trains" of a given length (see fig. 10.2) that can be made from Cuisenaire rods® (EDC 2003; Pagni 1998; Parker 1991). Many variations on the problem exist—using only rods of specified length, counting the number of "cars" of a specified size in the collection of trains of a given length, and many others. We have seen some very clever approaches and solutions to the problem. But all the solutions start with a representation of the context. Different representations suggest different approaches—finding algorithms to generate all the possibilities, lining up all the rods of length 1 to make the specified length and inserting dividers, making all the trains of length n from trains of length $n - 1$ in a systematic way, and a host of others. The mathematical habit of *representing*—mapping a new situation into one that is better understood (Cuoco 1991)—is ubiquitous in mathematics.

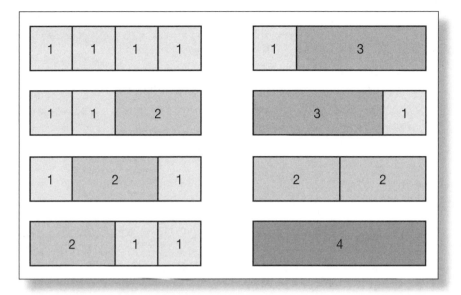

Fig. 10.2. Eight trains of length 4 can be made with colored rods.

Generalizing from Examples

Often students look at a problem and have no idea how to start. A common mantra many teachers use is, "Try it with numbers." Mathematics is open to experiment, and general results or at least conjectures often spring from trying specific examples, looking for regularity, and seeking what seem to be general trends. For example, we ask first-year algebra students to think about which positive integers can be represented as a *difference* of two perfect squares. This question can lead to a lovely result, but there is another reason to include it in an algebra course: It helps students develop the habit of generalizing from examples.

Articulating Generality in Precise Language

Another of our investigations asks students to consider the question, "Which integers can be written as the *sum* of two perfect squares?" Within twenty minutes or so, many students are able to tell whether various integers can or cannot be so written, and many can predict without actually trying all the possibilities. But it takes another half hour for a group to arrive at a consensus about how to describe its tests in precise mathematical language. An often-overlooked fact is that one can understand something without being able to articulate this understanding, and the habit of shoehorning insights into precise mathematical language takes time to develop. One device that we have found very effective in developing this skill is to include recurring dialogues among a small set of characters with different personalities and strengths—students who struggle with how to say what they know. By reading or acting out these dialogues, students begin to have similar conversations with their classmates, and they gradually develop this habit of articulating their ideas in precise language.

Expecting Mathematics to Make Sense

This is a central message of NCTM's *Focus in High School Mathematics: Reasoning and Sense Making* (2009a), which provides many examples from the high school curriculum. Another example comes from the algebra companion to that document, *Reasoning and Sense Making in Algebra* (NCTM 2009b): A triangle is determined by its three side lengths. It makes sense, then, that one should be able to find the area of a triangle from the lengths of its sides. Finding that area is not a simple task, but this insight leads one to believe that a *formula* for the area of a triangle in terms of its side lengths should exist. It does, and algebra can lead one to Heron's formula. Along these lines, a triangle is also determined by the lengths of its three medians. So there should be a formula for the area of a triangle in terms of these lengths. Readers are invited to find one.

Analytic and Geometric Habits of Mind

Certain habits of mind seem to be more prevalent in specific branches of mathematics. Thinking about continuous variation, dynamically changing systems, continuous deformations, and continuous functions are all much more common in geometry, analysis, and physics than they are in, say, algebra and combinatorics. That is not to say that algebra is devoid of considerations about change, but we have found that geometry and "precalculus" are much more suited to the development of this kind of thinking. Consider the following examples.

Reasoning by Continuity

The mathematician Thomas Banchoff asks his advanced calculus students on the first day of class, "Was there a time in your life when your height in inches was equal to your weight in pounds?" One approach is to try to find such a time. Another is to invoke continuity and an informal appeal to the thinking behind (and not necessarily the statement of) the intermediate value theorem of calculus.

Seeking Geometric Invariants

Hand in hand with the habit of reasoning about continuously changing systems is the habit of seeking *invariants* in those systems—things that *do not* change (Cuoco and Goldenberg 1997).

Looking at Extreme Cases and Passing to the Limit

This habit of mind is exemplified in the insight of a high school student we observed many years ago. On a standardized test, Rich was given an equilateral triangle of side length 10 and a point P somewhere in the triangle's interior. The problem was to find the sum of the distances from that point to the sides of the triangle. A theorem about this exists, but Rich did not know it. Instead, he reasoned that, because the question did not say otherwise, he could place P anywhere in the triangle's interior—he assumed that the sum of the distances from P to the sides of the triangle is constant. And then, in his mind, he tucked P into a corner of the triangle, where two of the distances go to zero and the other approaches the height of the triangle. Passing to the limit, he had a number he could calculate. Dynamic geometry allows Rich's thought experiment to become a real one, which is elaborated in Cuoco, Goldenberg, and Mark (1995). Again, the result—that the sum of the distances is constant—and the ensuing corollaries and generalizations are interesting in their own right, but another reason for putting this investigation into a high school program is that it supports the habit of looking to extremes.

Modeling Geometric Phenomena with Continuous Functions

One way to think about the insight shown in the previous example is to consider a function defined on the interior of a triangle whose value at a point is the sum of the distances to the sides of the triangle. The assumption that Rich made informally is that this function is constant on the triangle's interior. Continuity and analytic thinking, based in the structure of the real numbers, have always been just under the surface in plane geometry, going back to Euclid (Cuoco and Goldenberg 1997).

Algebraic Habits of Mind

Just as certain mathematical habits are indigenous to geometry and analysis, algebraists use certain ways of thinking quite often. These have to do with finding patterns in calculations, expressing regularity as algorithms, and looking for structural similarities among various systems in which one can calculate. Consider the following seven examples.

Seeking Regularity in Repeated Calculations

This habit manifests itself when one is performing the same calculation over and over and begins to notice the rhythm in the operations. Seeking and articulating this regularity is a backbone of algebraic thinking (Cuoco, Goldenberg, and Mark 2009). The following brief example is developed more fully in our Algebra 1 textbook (EDC 2009, p. 473):

> Suppose you want to buy a music CD, and a web site offers a 28% discount on the list price. It also adds a 5% state sales tax and a $3.50 shipping charge. The local music store sells CDs for 10% of list price, also charges 5% sales tax, but has no shipping charge. Ignoring convenience and drive time, for which list prices is it better to go online?

There are many ad hoc ways of solving such problems, but a core algebraic approach is to develop a function or algorithm that will allow one to compute the price of a CD in each scenario easily. For example, when buying from the Web site, one determines the total cost of a CD by taking the list price, subtracting 28% of that amount, adding 5% of that amount, and then adding $3.50. This may sound simple enough, but many students are not able to articulate this process without referring to a specific price, like $25. In other words, they can compute the total cost of a CD for any particular list price but cannot express the general process for calculating the total cost of a CD. And expressing that process is precisely the first step of what is needed to find the break-even point. One way to develop the skill of expressing this kind of generality is to begin with some numerical examples until one gets the rhythm of the calculations and is able to articulate the cost of a CD in a way that will work for any price.

Delayed Evaluation—Seeking Form in Calculations

General patterns often get masked when expressions or numerical calculations are evaluated too early. Often in algebra one wants to delay numerical evaluation until the end of a process so that one can see how the operations are sequenced and so that the structure of the calculation (rather than its value) becomes more apparent. For more about delayed evaluation, see NCTM (2009b) and Cuoco (2005).

Chunking—Changing Variables to Hide Complexity

Often in algebra one wants to treat a whole expression as a single object. Teachers often cover up parts of an expression with a hand and ask students to think of what is under the hand as a single entity. For example, $x^4 + 2x^2 + 1$ can be considered as a quadratic in x^2:

$$x^4 + 2x^2 + 1 = (x^2)^2 + 2(x^2) + 1 = \clubsuit^2 + 2\clubsuit + 1$$

Chunking has applications throughout high school algebra. Further examples are described in Cuoco (2008) and NCTM (2009b).

Reasoning about and Picturing Calculations and Operations

A key habit of mind in algebra involves predicting how a calculation will go without having to carry it out. For example, consider the following question: "When is the average of two averages the average of the whole lot?" After a little experimentation, one begins to reason about the *process* of averaging and comes up with several situations in which the statement is true and several more when it is not. More examples of this habit are explicated in *Focus in High School Mathematics: Reasoning and Sense Making* (NCTM 2009a) and in the *College- and Career-Readiness Standards* (Common Core State Standards Initiative 2009).

Extending Operations to Preserve Rules for Calculating

Suppose we extend the multiplication table shown in figure 10.1 to the other quadrants (see fig. 10.3). How could we fill in the blanks to maintain patterns in the rows and columns? There are many answers to this question because there are many ways to describe patterns in the rows and columns. But, in fact, there is only *one* way to complete the table that will preserve the basic rules for calculating with positive integers (e.g., the distributive law).

Curriculum Issues in an Era of Common Core State Standards for Mathematics

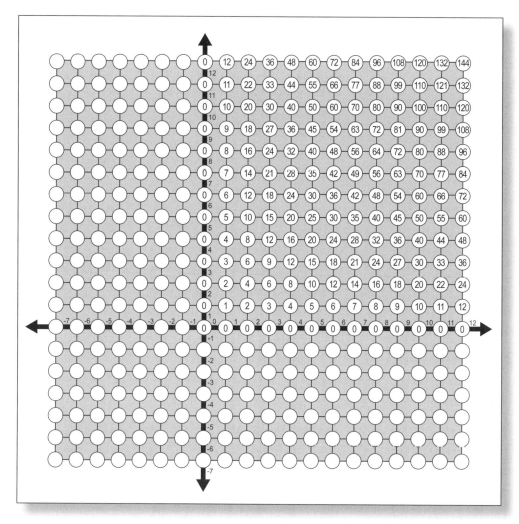

Fig. 10.3. How can the blanks be filled in so as to preserve the pattern and the calculation rules?

This habit of extending to preserve rules is a key algebraic habit of mind. The point is that extensions of algebraic operations are not arbitrary—they are forced on us by the desire to maintain the basic rules of arithmetic.

Purposefully Transforming and Interpreting Expressions

Focus in High School Mathematics: Reasoning and Sense Making (NCTM 2009a) uses the phrase *mindful manipulation* when it refers to the habit of transforming algebraic expressions to reveal hidden meaning. Writing $3x^2 - 12x + 16$ as $3(x-2)^2 + 4$ tells one that, when x is replaced by any real number, the value of the expression is never smaller than 4. Writing $15x^2 - 57x + 46$ as

$$5(x-1)(x-2) + 2(x-2)(x-3) + 8(x-1)(x-3)$$

118

makes it easy to evaluate the quadratic at 1, 2, and 3. Writing Heron's formula in the form

$$A = \frac{1}{4}\sqrt{(a+b+c)(a+b-c)(a+c-b)(b+c-a)}$$

helps one see exactly when the formula produces zero (NCTM 2009b). Finding examples where purposeful transformations produce added value and utility is an important part of our curriculum development efforts.

Seeking and Specifying Structural Similarities

Watch high school juniors and seniors calculate with complex numbers in the form $a + bi$. Many act as if i is x and complex numbers are just polynomials. Somewhere in the calculation, they replace i^2 with -1. Without using any formal language, these students are noticing that the system of complex numbers behaves like the system of polynomials in one variable with real coefficients, with an extra simplification rule ($i^2 = -1$). Students are noticing a close similarity between the algebraic structures of the complex numbers and polynomials over the real numbers.

Developing the habit of looking for and describing these structural similarities is one of the basic habits in modern algebra, and the high school curriculum provides many opportunities to highlight it. Just a few examples: Factoring polynomials is quite a bit like factoring integers; 2 × 2 matrices and their multiplication have the same structure as linear transformations of the plane under composition; and roots of unity behave like polynomials with a certain remainder arithmetic.

Conclusion

One of the most satisfying aspects of this habits-of-mind focus is that the authors were able to connect many seemingly different topics from the high school curriculum by looking beneath the topics to find approaches that revealed remarkable conceptual similarities. This approach not only brought coherence to the program; it also brought parsimony. By focusing on the underlying habits of mind needed to deal with the topics, we were able to drastically reduce the method bloat that plagues school mathematics, providing students with a few general-purpose tools that could be applied across the entire curriculum (Cuoco 2008).

Space prohibits their inclusion here, but mathematical habits of mind allow one to bring coherence to many different parts of the curriculum; see Cuoco, Goldenberg, and Mark (2009).

When we began this line of work—organizing curricula around mathematical thinking—our goal was to make school mathematics both more accessible to high school students and more closely aligned with mathematics as a scientific discipline. Once implemented, the method had other benefits—coherence among topics and parsimony of methods. In hindsight, these benefits and results should have been obvious. The results and methods of mathematics are its artifacts; the actual mathematics lies in the thinking that creates the artifacts.

References

Cobb, George W., and David S. Moore. "Mathematics, Statistics, and Teaching." *American Mathematical Monthly* 104 (November 1997): 801–23.

Common Core State Standards Initiative. *College- and Career-Readiness Standards*. 2009. http://www.corestandards.org.

Cuoco, Al, ed. *The Roles of Representation in School Mathematics*. 1991 Yearbook of the National Council of Teachers of Mathematics (NCTM). Reston, Va.: NCTM, 1991.

―――. *Mathematical Connections: A Companion for Teachers and Others*. Washington, D.C.: Mathematical Association of America, 2005.

―――. "Introducing Extensible Tools in Middle- and High-School Algebra." In *Algebra and Algebraic Thinking in School Mathematics*, 2008 Yearbook of the National Council of Teachers of Mathematics (NCTM), edited by Carole E. Greenes, pp. 51–62. Reston, Va.: NCTM, 2008.

Cuoco, Al, and E. Paul Goldenberg. "Dynamic Geometry as a Bridge from Euclidean Geometry to Analysis." In *Geometry Turned On: Dynamic Software in Learning, Teaching, and Research*, MAA Notes 41, edited by James Richard King and Doris Schattschneider, pp. 33–44. Washington, D.C.: Mathematical Association of America, 1997.

Cuoco, Al, E. Paul Goldenberg, and June Mark. "Connecting Geometry with the Rest of Mathematics." In *Connecting Mathematics across the Curriculum*, 1995 Yearbook of the National Council of Teachers of Mathematics (NCTM), edited by Peggy A. House, pp. 183–97. Reston, Va.: NCTM, 1995.

―――. "Habits of Mind: An Organizing Principle for Mathematics Curricula." *Journal of Mathematical Behavior* 15 (December 1996): 375–402.

―――. "Curricular Coherence through a Focus on Mathematical Thinking." 2009. http://www2.edc.org/cme/HOM/hom-high-school.pdf.

Education Development Center Inc. (EDC). "Warm Ups for the Simplex Lock Problem." Making Mathematics, 2003. http://www2.edc.org/makingmath/mathprojects/simplex/simplex_warmup.asp.

―――. *The CME Project: Algebra 1, Geometry, Algebra 2, Precalculus*. Boston: Pearson Education, 2009.

Mark, June, Al Cuoco, E. Paul Goldenberg, and Sarah Sword. "Developing Mathematical Habits of Mind in the Middle Grades." Working paper. Education Development Center, 2009.

National Council of Teachers of Mathematics (NCTM). *Curriculum and Evaluation Standards for School Mathematics*. Reston, Va.: NCTM, 1989.

―――. *Focus in High School Mathematics: Reasoning and Sense Making*. Reston, Va.: NCTM, 2009a.

―――. *Reasoning and Sense Making in Algebra*. Reston, Va.: NCTM, 2009b.

Pagni, David. "Cuisenaire Trains." *Mathematics in School* 27 (January 1998): 16–17.

Parker, Ruth E. "Implementing the *Curriculum and Evaluation Standards*: What Will Implementation Take?" *Mathematics Teacher* 84, no. 6 (September 1991): 442–49, 478.

Scheaffer, Richard, and Josh Tabor. "Statistics in the High School Mathematics Curriculum: Building Sound Reasoning under Uncertain Conditions." In *Curriculum Issues in an Era of Common Core State Standards for Mathematics*, edited by Christian R. Hirsch, Glenda T. Lappan, and Barbara J. Reys, pp. 185–93. Reston, Va.: National Council of Teachers of Mathematics, 2012.

Thurston, William P. "On Proof and Progress in Mathematics." *Bulletin of the American Mathematical Society* 30 (April 1994): 161–77.

Section V
Selecting and Strategically Using Technology Tools and Resources to Support Mathematics Teaching and Learning

Introduction

Technology is essential in teaching and learning of mathematics; it influences the mathematics that is taught and enhances student learning. (National Council of Teachers of Mathematics 2000, p. 24)

In Section IV, you examined the Standards for Mathematical Practice of the Common Core State Standards for Mathematics (CCSSM), how they are related to previous work with mathematical habits of mind, and steps toward making the mathematical practices a focus of your grades K–12 mathematics program. In this section, you will examine more closely the intent and instructional implications of the mathematical practice *use appropriate tools strategically* (Common Core State Standards Initiative 2010, p. 7). The three related chapters in Section V collectively focus on issues relevant to improving the use of technology tools in classrooms; the support needed for teachers to use calculators in interesting, productive ways; examples of ways that existing and emerging technologies can encourage, support, and enhance students' engagement with and learning of mathematics; and the effect of computer algebra systems (CAS) on learning the big ideas in algebra.

Each chapter explores effective, grade-appropriate uses of technology tools in mathematics classrooms. Knowing what experiences with technology students may bring to your grade, and what the expectations are for subsequent grades, can help you build smoother transitions for your students. The chapter by Chval and Hicks gives examples of work on mathematical tasks from the elementary school mathematics curriculum that illustrate how you can use calculators both appropriately and strategically. The chapter by Hollenbeck and Fey explores how students' and teachers' full use of existing mathematical and communication tools could change how you teach mathematics in the middle grades and assess students' learning, as well as how it would change content of your curriculum. Finally, the chapter by Zbiek and Heid explores features of CAS, particularly when used as part of a classroom tool set, and addresses several ways in which CAS can focus the taught and learned curriculum on big mathematical ideas that might have previously escaped students' attention. Zbiek and Heid also list examples of how CAS use supports several CCSSM mathematical practices.

As you read these three chapters, think about how they relate to the *using appropriate technology tools strategically* mathematical practice and what they imply for you as a practicing teacher and for your school's and district's mathematics program.

Questions for Reflection and Collective Discussion

1. What did you find particularly interesting or noteworthy in your reading from this set of three chapters? Be prepared to discuss those ideas with colleagues. How, if at all, can you incorporate those ideas into your instructional practice?

2. What policies and resources does your district have in place that support teachers' and students' use of technology tools in grades K–5? In grades 6–8? In grades 9–12? Do your readings reinforce or challenge those policies?

3. What challenges do you face in using technology, especially calculators and computer tools, to support students' learning?

4. From your perspective, what role do curriculum materials play in supporting students' development of the CCSSM mathematical practice *use appropriate technology tools strategically*? What curriculum design features might be most helpful?

5. How can you address the equity issue related to giving all students access to technology tools in the classroom? Outside the classroom?

6. How do you use calculators and computer tools effectively to improve students' learning and support problem solving in your classroom?

7. How can you help students become strategic about when to use technology tools and when not to? How can you reinforce this important habit across grades K–12?

8. Visit the NCTM Illuminations website (http://www.illuminations.nctm.org) and, if possible, identify and use one of the instructional applets in your classroom. Discuss your experience with colleagues from the perspectives of teaching and students' engagement and learning.

9. In the words of the CCSSM document, "Mathematically proficient students consider the available tools when solving a mathematical problem. These tools might include pencil and paper, concrete models, a ruler, a protractor, a calculator, a spreadsheet, a computer algebra system (CAS), a statistical package, or dynamic geometry software." Which of these tools are appropriate for the class(es) you teach? Which of these tools do your students currently have available and use? If needed, what steps could your school or district take to ensure students access to appropriate technology in learning and doing mathematics?

10. Scan the CCSSM content standards for the grade band—K–5, 6–8, or 9–12—at which you teach. Which content standards, if any, specifically refer to technology use? For those that do, identify the technology tool and how it is intended to be used. For the remaining standards, identify standards for which you think strategic technology use would be appropriate. Discuss your findings with your colleagues, paying special attention to differences in perspective.

11. Look back at the chapter(s) you read in this section and identify specific mathematical practices that they exemplify. Compare your findings with those of your colleagues.

12. Using calculators and computer tools can affect the curriculum in at least three ways:

    a. Some mathematics becomes *more important* because technology *requires* it.

    b. Some mathematics becomes *less important* because technology *replaces* it.

    c. Some mathematics becomes possible because technology *allows* it.

    How can you see these effects today in your school mathematics curriculum in grades K–5? In grades 6–8? In grades 9–12? How can you see these effects in the CCSSM?

13. Public domain computer algebra systems (CAS) are freely available at the following websites:
    - www.wmich.edu/cpmp/CPMP-Tools
    - www.sourceforge.net/projects/xcas
    - www.wolframalpha.com

    Organize study groups, and then select and experiment with different software packages. Report to colleagues which software features you think would help implement the CCSSM content and mathematical practices standards.

References

Common Core State Standards Initiative. *Common Core State Standards for Mathematics.* Washington, D.C.: National Governors Association Center for Best Practices and the Council of Chief State School Officers, 2010. http://www.corestandards.org.

National Council of Teachers of Mathematics (NCTM). *Principles and Standards for School Mathematics.* Reston, Va.: NCTM, 2000.

Chapter 11

Strategically Using Calculators in the Elementary Grades

Kathryn B. Chval
Sarah J. Hicks

For many years, the National Council of Teachers of Mathematics (NCTM) and other national organizations and prestigious committees have recommended the use of calculators in the classroom (National Council of Supervisors of Mathematics [NCSM] 1988; NCTM 1989, 2000; National Research Council [NRC] 1990). NCTM again emphasized the importance of calculators in 2008 with its release of a position statement on the use of technology, arguing that curricula should incorporate instructional technology into learning outcomes, lesson plans, and assessments of students' progress and explain why the use of calculators and other technological tools is important (NCTM 2008). Standards documents have also advocated for the use of calculators at both the state and national levels, although at different levels of priority and prominence.

Calculators have the potential to influence both mathematics teaching and learning positively. Yet, this influence requires the availability of calculators, teachers who know how to use calculators as well as how to teach with them effectively, curriculum materials that facilitate their use, and supportive school and community environments. Unfortunately, too often, these conditions do not exist in U.S. elementary schools. In this chapter, we highlight how curriculum standards documents, including the Common Core State Standards for Mathematics (CCSSM; Common Core State Standards Initiative [CCSSI] 2010), treat calculator use and we present examples of mathematical tasks from elementary school mathematics curricula that provide children opportunities to use calculators strategically. In addition, we describe how elementary school teachers can cultivate the CCSSM mathematical practice of strategically using calculators as expected by the CCSSM.

Calculators in Mathematics Curriculum Standards

Prior to the introduction of the CCSSM, most states generated curriculum standards with specific grade-level expectations (GLEs) for mathematics in elementary school. An analysis of these state standards documents noted that 36 of 43 state documents included a discussion of calculators either in introductory material (20 state documents) or in statements of specific learning expectations (23 state documents) (Chval, Reys, and Teuscher 2006). Across these 23 state documents, 186 GLEs referred to calculator use in grades K–5. State standards described different uses of calculators for students (see table

11.1 for different calculator roles). These standards go beyond simply expecting to use a calculator to compute with larger numbers or to check answers of computations solved using paper and pencil.

Table 11.1
Roles of Calculators in GLEs in State Curriculum Documents

Role of Calculator	Description
Represent	Students use calculators to represent mathematical quantities and ideas including different notations and graphs. They also connect physical models to mathematical language.
Solve problems or equations	Students use calculators to solve applied problems or equations.
Develop or demonstrate conceptual understanding	Students use calculators to build conceptual knowledge of mathematical ideas and demonstrate understanding of these concepts.
Analyze	Students use calculators to compare, interpret, and identify relationships; make predictions; or make sense of data.
Compute or estimate	Students use calculators to compute or estimate.
Describe, explain, justify, or reason	Students use calculators to help them describe strategies, explain reasoning, or justify mathematical thinking.
Choose appropriate method of calculation	Students determine whether they should compute mentally, use calculators, or use paper and pencil.
Determine the reasonableness of a calculated answer	Students determine whether the answer acquired using a calculator is reasonable.

Recently, the National Governors Association Center for Best Practices (NGA Center) and the Council of Chief State School Officers crafted and launched the CCSSM, which most states have formally adopted. The CCSSM is divided into two sections at the elementary school level: Standards for Mathematical Content and Standards for Mathematical Practice. The Standards for Mathematical Content do not refer to calculator use in elementary school. However, one of the eight Standards for Mathematical Practice refers to calculators as follows:

> Mathematically proficient students consider the available tools when solving a mathematical problem. These tools might include pencil and paper, concrete models, a ruler, a protractor, a calculator, a spreadsheet, a computer algebra system, a statistical package, or dynamic geometry software. Proficient students are sufficiently familiar with tools appropriate for their grade or course to make sound decisions about when each of these tools might be helpful, recognizing both the insight to be gained and their limitations.... They detect possible errors by strategically using estimation and other mathematical knowledge.

When making mathematical models, they know that technology can enable them to visualize the results of varying assumptions, explore consequences, and compare predictions with data.... They are able to use technological tools to explore and deepen their understanding of concepts. (CCSSI 2010, p. 7)

Similar to the roles of calculators represented in the state standards documents presented in table 11.1, the expected use of calculators in the CCSSM goes beyond their superficial use, such as computing or checking answers to mathematical problems.

Calculators in Curriculum Materials

Although NCTM and others advocate using calculators, their inclusion in elementary school mathematics curriculum materials is more sporadic. To investigate the extent and nature of calculator use in curriculum materials, we examined fourth-grade student and teacher materials from six curriculum series: Math Expressions (Houghton Mifflin 2006); Think Math! (Harcourt School Publishers 2008); Everyday Mathematics, 3rd edition (Wright Group/McGraw-Hill 2007); Math (Macmillan/McGraw-Hill 2005); Math Trailblazers, 2nd edition (Kendall/Hunt 2004); and Investigations in Number, Data, and Space, 2nd edition (Pearson Education 2008).

To determine the extent of use, we identified each page that referred to calculators in both the teachers' and students' materials. Because calculators were included in the lists of necessary materials for specific lessons but not for others, only pages referring to specific calculator use were counted. The authors of two series explicitly stated that calculators could not be used on some specific student pages. Student pages that prohibited calculator use were not included, because the study focused on when and how calculators were to be used in the materials. Table 11.2 summarizes the number of pages that refer to calculators for each curriculum series examined.

Table 11.2
Number of Pages Referring to Calculators in Students' and Teachers' Materials

Curriculum Series	Number of Pages in Students' Materials	Number of Pages in Teachers' Materials
Math Trailblazers	56	153
Everyday Mathematics	72	116
Macmillan/McGraw-Hill Math	46	9
Investigations	3	43
Math Expressions	1	8
Think Math!	4	5

As shown in table 11.2, the extent of references to calculators varied considerably across the six curriculum programs. Table 11.3 lists the percentage of daily lessons that mention calculators, in either the students' or teachers' materials. As shown in the table, some calculator use is indicated as optional in the teachers' materials.

Table 11.3
Percentage of Daily Lessons Involving Calculators

Curriculum Series	Percentage of Daily Lessons Calculator Use Is Expected	Percentage of Daily Lessons Indicating Optional Calculator Use
Math Trailblazers	62/160 = 39%	0%
Everyday Mathematics	27/126 = 21%	3/126 = 2%
Macmillan/McGraw-Hill Math	9/184 = 5%	14/184 = 8%
Investigations	2/160 = 1%	36/160 = 23%
Math Expressions	5/160 = 3%	3/160 = 2%
Think Math!	3/155 = 2%	0%

Note that students do not typically use calculators for an entire class period, so the percentages listed in the table should not be interpreted as a percentage of *time* that students use calculators, but rather as a percentage of daily lessons that involve calculator use at some level. For example, in a one-hour lesson students may use the calculator for only one computation.

In addition to variation in the extent of calculator use, dramatic differences exist related to the nature of calculator use across the materials. For example, note the following:

- Everyday Mathematics and Macmillan/McGraw-Hill Math identify specific calculator models, whereas the other curricula do not. Everyday Mathematics includes directions for using both the TI-15 calculator and the Casio fx-55 calculator. Macmillan/McGraw-Hill Math uses the TI-15 calculator.
- Macmillan/McGraw-Hill Math, Math Trailblazers, and Everyday Mathematics include lessons on how to use specific calculator features.
- Both Math Trailblazers and Everyday Mathematics indicate when calculators should be used on assessments; the others do not.
- The teacher materials for Math Trailblazers, Everyday Mathematics, and Investigations indicate specific times when calculators should *not* be used.
- Everyday Mathematics and Math Trailblazers have extensive notes in the teachers' materials on how to use calculators and potential mathematical challenges that may occur in class discussions (e.g., the difference between the subtract key and the negative key). In letters to parents and caregivers, Everyday Mathematics and Math Trailblazers also discuss calculator use.

- Math Trailblazers and Everyday Mathematics include mathematical problems that address the eight calculator roles identified in the GLEs (see table 11.1).
- Across the six curriculum series, calculator use is described in different features of the materials such as enrichment activities, technology links, readiness activities, homework assignments, assessments, and classroom routines.

To illustrate the use of calculators in specific curriculum features, figure 11.1 compares Everyday Mathematics and Math Expressions. Note, not all the instances in which curriculum materials use calculators require appropriate uses of calculators.

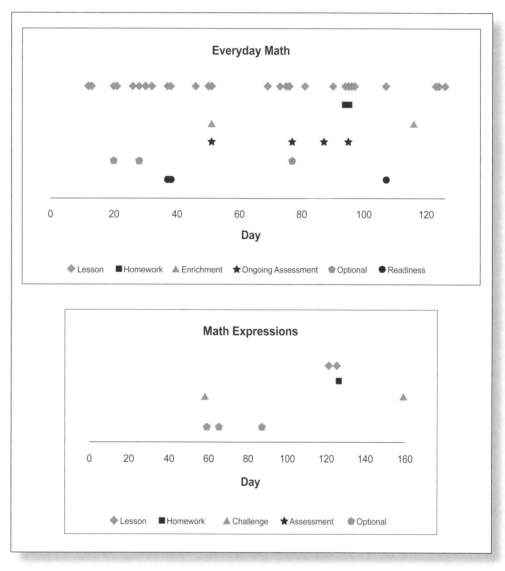

Fig. 11.1. Days that involve calculator use for Everyday Math and Math Expressions

However, we present a few mathematical tasks from elementary school mathematics curriculum materials to illustrate problems that require appropriate and strategic use of calculators. Think Math! has minimal references to calculators; however, figure 11.2 illustrates one example in the student materials from that series.

Use a calculator to multiply these numbers between 6.5 and 6.6.

Numbers Between 6.5 and 6.6 →	Numbers Multiplied by Themselves
6.5	
6.51	
6.52	
6.6	

Paul multiplied a number by itself and got 43. His number must be between two numbers in the left column.

His number must be between _____ and _____.
How do you know?
_____
_____

Challenge Name three numbers between 3.65 and 3.66.

☐ ☐ ☐

Fig. 11.2. Sample problem involving calculators from Think Math! Lesson Activity Book Grade 4 (2008, p. 154)

The *Everyday Mathematics Teachers Reference Manual* (2007, p. 6) states, "Special emphasis is placed on using calculators as tools for counting, displaying numbers, developing concepts and skills, and solving problems—especially real-life problems in which numbers are not always 'nice.'" The problem in figure 11.3 illustrates how fourth graders use calculators in Everyday Mathematics to work with numbers from real-life data.

> In 2001, there were about 2,317,000 marriages in the United States. The table below shows the approximate number of marriages each month.
>
> 1. Use a calculator to find the percent of the total number of marriages that occured each month. Round the answers to nearest whole-number percent.
>
> | Month | Approximate Number of Marriages | Approximate Percent of Total Marriages |
> |---|---|---|
> | January | 147,000 | 6% |
> | February | 159,000 | |
> | March | 166,000 | |
> | April | 166,000 | |
> | May | 189,000 | |
> | June | 237,000 | |
> | July | 244,000 | |
> | August | 225,000 | |
> | September | 224,000 | |
> | October | 217,000 | |
> | November | 191,000 | |
> | December | 152,000 | |
>
> *Source:* U.S. Department of Health and Human Services
>
> 2. According to the table, what is the most popular month for a wedding? _____
>
>   What is the least popular month? _____
>
> 3. Describe how you used your calculator to find the percent for each month.
>   _____
>   _____

Fig. 11.3. Sample problem from Everyday Mathematics Math Masters Grade 4 (2007, p. 288)

Figure 11.4 illustrates one example from Math Trailblazers where students use calculators to build understanding for number concepts.

Figures 11.2–11.4 show that current elementary school mathematics curricula include creative, challenging mathematical tasks that use the calculator as a tool. In these instances, calculators help students investigate and solve problems that would be impossible or too time-consuming to solve without a calculator. The calculator creates new possibilities for mathematical tasks that were not plausible before the introduction of calculators. Moreover, "the use of realistic data is motivational and helps children see connections between school mathematics and mathematics used in the world" (Charles 1999, p. 11).

Curriculum Issues in an Era of Common Core State Standards for Mathematics

Operation Target

This is a cooperative contest for two or three people. The goal is to use four digits and four operations (+, –, ×, and ÷) to make as many different whole numbers as you can. You need paper, pencil, and a calculator.

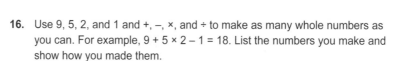

You must use each of the four digits exactly once. You can use operations more than once or not at all. (All division operations must give whole numbers. For example, 9 ÷ 2 = 4.5 is not allowed.)

16. Use 9, 5, 2, and 1 and +, –, ×, and ÷ to make as many whole numbers as you can. For example, 9 + 5 × 2 – 1 = 18. List the numbers you make and show how you made them.

    A. What is the largest whole number you can make?

    B. What is the smallest whole number you can make?

    C. How many whole numbers less than 10 can you make? Write number sentences for each number.

    D. What whole numbers can you make in more than one way? Show at least two number sentences for each.

17. Pick four different digits. Make as many whole numbers as you can using your four new digits and +, –, ×, and ÷. List the numbers you make and show how you made them.

18. Nila used 1, 2, 3, and 4 to make 10. How do you think she did it? Can you think of another way?

19. Luis use 1, 2, 3, and 4 to make 24. How could he have done it?

20. Romesh used 1, 3, 5, and 7 to make 8. How could he have done it?

21. Make up your own problem like Questions 18, 19, and 20.

Fig. 11.4. Sample problem from *MathTrailblazers Student Guide Grade 4* (2004, p. 183)

Although current elementary school mathematics curriculum materials cite many creative and potentially effective uses of calculators, some also present problematic aspects of calculator use:

- Calculator activities are "add on" or "optional." They are primarily used in "extra features," such as extension problems, rather than embedded in lessons.
- Calculators are used solely for checking answers.

- Students are "exposed" to calculators rather than having opportunities to use them consistently and strategically.
- Students use the calculator for a single word problem that has more complicated numbers.
- Calculators are used unnecessarily (e.g., 77 – 65).
- Materials show an example of how to use calculators but then do not offer opportunities for students to use calculators themselves.
- Materials create limited opportunities for students to develop conceptual understanding; analyze situations; or describe, justify, and reason with calculators.
- Teachers' materials do not give teachers sufficient guidance to use calculators effectively and promote their strategic use.

Such aspects of calculator use are problematic in that teaching demands often constrain the use of activities that teachers view as optional or unnecessary. When calculators are used rarely or only for one computation in a lesson, distributing and managing calculators is not worth valuable class time. When activities are labeled as "optional," teachers will probably skip them so they can address required material. Similarly, when teachers have not had sufficient opportunities to learn how to use the calculator's memory system, trial-and-error feature, or fraction keys, they cannot introduce these features to students. Moreover, if students do not engage while using calculators, then teachers will be less likely to use them. Finally, teachers do not have opportunities to realize the calculator's potential as a teaching and learning tool when curriculum materials do not give students opportunities to develop conceptual understanding; analyze situations; or describe, justify, and reason with calculators.

Most elementary school curriculum materials on the current market were developed prior to the introduction of the CCSSM. As a result, in most instances they do not support teachers' cultivating the mathematical practice of strategically using calculators. Therefore, most elementary school teachers need to consider how to enhance curriculum materials so that their students learn how to use calculators appropriately and strategically.

Calculators in Classrooms

With guidance from effective mathematics teachers, students at different levels can use calculators to support and extend mathematical reasoning and sense making, gain access to mathematical content and problem-solving contexts, and enhance computational fluency. In a well-articulated mathematics program, students can use calculators for computation, construction, and representation as they explore problems. Using calculators also contributes to mathematical reflection, problem identification, and decision making. To describe how elementary school teachers can cultivate the mathematical practice of strategically using calculators, we offer examples from one teacher.

Sara, a fifth-grade teacher, taught mathematics in a low-income, urban community, using Math Trailblazers as well as supplementary curriculum materials that required calculator use. When her students entered her classroom, they answered her questions with numbers or short phrases. They did not know how to work productively in groups. They did not exhibit the Standards for Mathematical Practice identified in the CCSSM:

- Make sense of problems and persevere in solving them.
- Reason abstractly and quantitatively.
- Construct viable arguments and critique the reasoning of others.
- Model with mathematics.
- Use appropriate tools strategically.
- Attend to precision.
- Look for and make use of structure.
- Look for and express regularity in repeated reasoning.

Yet, by second semester, these mathematical practices, including the strategic use of calculators, were the norm in Sara's classroom. Sara introduced the term *keystrokes* in the context of a specific machine: the scientific calculator. Sara used this term in two distinct ways: (1) to denote striking a calculator key and (2) to speak or write the symbol representing the calculator key. Sara required students to present keystrokes both orally and in writing so that the students could use keystrokes to communicate and discuss mathematical thinking. She made students responsible for articulating reasoning and for working hard to understand the reasoning of others, crucial elements of the CCSSM mathematical practices and NCTM's (2000) Process Standards.

In Sara's classroom, using keystrokes to discuss and negotiate meaning was more important than using the calculator to compute answers. The following excerpts from Sara's lessons demonstrate what she said to convey the importance of the keystrokes.

Sara: On your paper, write the keystrokes that you would need to put into your calculator to find the area of that rectangle. Don't take the calculators out; I didn't say that. I asked you to write the keystrokes.

Sara: Your keystrokes are very important to me, because they tell me what you are thinking. I cannot be inside your head. Oh, unless I open his head. [class laughter] I can't do it. But if I see your work, I know what you are thinking.

Sara used calculator keystrokes as a common language, central to the classroom discourse, to mediate interactions in small-group and whole-class discussions. Students presented keystrokes both orally and in writing to communicate mathematical thinking, initiated conversations and posed questions to their peers and teacher by using this language, and analyzed their peers' solutions and articulated corrections and modifications. The students used the keystrokes to plan and create strategies to solve mathematics problems, to analyze others' strategies, and to create and articulate a process of reasoning to justify their solutions. In fact, Sara's students became so proficient at thinking and problem solving with the keystrokes that they no longer needed the calculators themselves. Figure 11.5 outlines Sara's technique for using keystrokes in a typical mathematics problem. Note that Sara never explicitly articulated this technique, but instead used the process, except for step 6, in every observed lesson. (For more discussion of Sara's use of writing in the mathematics classroom, see Chval and Khisty 2009.)

Step 1	Students write a plan for solving the problem, using only keystrokes.
Step 2	Students use keystrokes to communicate the plan.
Step 3	Students break down the sequence of keystrokes into components and discuss the meaning of each component.
Step 4	Students listen to a presentation of keystrokes, analyze keystrokes, and make a decision concerning agreement with that solution.
Step 5	Students suggest alternative keystroke sequences, including more efficient ones.
Step 6	Students write a narrative of how to solve the problem.

Fig. 11.5. Sarah's process for using calculator strokes

Through this process, Sara used keystrokes to facilitate developing the functions of planning, problem solving, reflection, analysis, and writing. For example, Sara used calculator keystrokes to develop the students' "planning function," or their ability to plan for the solution of a problem. She expected students not only to write their keystrokes but also to write them before they touched the calculators. In an important variation from common teaching practice, Sara did not use keystrokes only as a record of what a child had already pressed. The keystrokes, as used in Sara's classroom, served to communicate and display students' thinking but, more important, served to create a specific plan for solving a problem before the students touched the calculators.

Once the students developed a plan for solving a problem, Sara used the keystrokes as referents for discussing mathematical ideas. As students presented sequences of keystrokes to the class, they were challenged to communicate their thinking by writing the keystrokes on the chalkboard and orally explaining the solution. Sara used this public process to help students clarify their thinking and build meaning. She posed questions such as the following:

- What does 6 represent?
- Why do you press the multiplication symbol?
- Why did you divide?
- What does the 6 times 8 equals represent?
- Where did the 2 come from? I don't see a 2 on the picture.

Sara's questions communicate to the students that the explanation for each keystroke is important. Sara asks questions so that students would explain the purpose for using each keystroke and the meanings behind the values that appear during the process. As two students presented at the chalkboard, Sara explained that it was the responsibility of all members of the class to listen to the presentation, analyze the keystrokes, and decide whether they agreed or disagreed with the keystrokes. Even if a student had written a different sequence on his or her paper, he or she was responsible for understanding other solutions presented by peers. This scope included recognizing invalid methods, suggesting alternative methods, or determining more efficient methods.

Disagreement or argumentation can be established as a negative, confrontational practice; Sara, however, carefully created an environment whereby disagreement was cooperative, positive, and valued. Interestingly, she used disagreement as a mechanism for developing processes such as reflection, analysis, reasoning, and justification. She situated students and herself in a position to "construct viable arguments and critique the reasoning of others" (CCSSM Mathematical Practice 3). Sara established the importance of stating disagreement when it was observed. She also explicitly explained that disagreements warrant discussion. Although some educators may think that Sara's emphasis on keystrokes reinforced an algorithmic way of solving problems, an analysis of students' work indicated students often generated five to eight different solution strategies for many of the problems. This finding strongly suggests that students were not applying memorized procedures for solving problems, but rather understanding why mathematical statements were true—a goal the CCSSM identifies. Thus, Sara gives one example of how a teacher could use calculator keystrokes to facilitate mathematical thinking and communication. She shows how a teacher can integrate calculator use with mathematics curriculum materials to enact the CCSSM mathematical practices.

Recommendations

The presence of calculators in grades K–5 classrooms and mathematics curriculum materials was rare thirty years ago, and resistance to calculator use was strong and widespread. Since then, calculators have become more sophisticated and cost-effective as the resistance to their use in classrooms has decreased (Chval 2008). Today, state and national standards, curriculum materials, and assessments make reference to calculators. All these indicators suggest that calculator use in grades K–5 classrooms is more common and acceptable. However, as discussed above, indications are that further improvement is warranted to use the potential of calculators best in K–5 mathematics classrooms. The question still applies: "How should we make use of this extraordinary technology to further the mathematics education of our students?" (Usiskin 1999, p. 1).

New CCSSM-oriented curriculum materials should avoid the problematic issues of some current materials on the market, and more purposeful, consistent use of calculators should be incorporated within daily lessons. Also, teacher materials must provide sufficient guidance on how to use the calculator and how to use the calculator to teach children. NCTM's 2008 position statement cautions that "teacher education and professional development must continually update practitioners' knowledge of technology and its classroom applications."

Although the CCSSM document includes the strategic use of calculators, it does not offer specific advice about how to develop this mathematical practice for specific grades. Therefore, teachers will need more support to cultivate this practice. As calculator technology changes, teachers need opportunities to learn not only how to use the tools well but also how to teach with them effectively. Regardless of the curriculum materials or calculator model that a specific district uses, teachers need print resources, professional development, and a supportive environment to facilitate learning how to use calculators appropriately and strategically, especially as calculator technology improves and grades K–5 curriculum materials take more advantage of calculators' potential. School or district leaders need to facilitate conversations and professional development related to using calculators to ensure effective schoolwide implementation.

References

Charles, Randall I. "Calculators at the Elementary School Level? Yes, It Just Makes Sense." *Mathematics Education Dialogues* 2 (May/June 1999): 11.

Chval, Kathryn B. "The Status of Calculator Technology in U.S. K–8 Mathematics Curriculum: It Depends How You Look at It." In *Mathematics Education in Pacific Rim Countries: China, Japan, Korea, and Singapore,* edited by Zalman Usiskin and Edwin Willmore, pp. 305–16. Charlotte, N.C.: Information Age Publishing, Inc., 2008.

Chval, Kathryn B., and Lena L. Khisty. "Latino Students, Writing, and Mathematics: A Case Study of Successful Teaching and Learning." In *Multilingualism in Mathematics Classrooms: Global Perspectives,* edited by Richard Barwell, pp. 128–44. Clevedon, U.K.: Multilingual Matters, 2009.

Chval, Kathryn, Barbara J. Reys, and Dawn Teuscher. "What Is the Focus and Emphasis on Calculators in State-Level K–8 Mathematics Curriculum Standards Documents?" *Mathematics Education Leadership Journal* 9 (Spring 2006): 3–13.

Common Core State Standards Initiative (CCSSI). *Common Core State Standards for Mathematics.* Washington, D.C.: National Governors Association Center for Best Practices and the Council of Chief State School Officers, 2010. http://www.corestandards.org.

Everyday Mathematics Grade 4. Everyday Mathematics Series. Chicago: Wright Group/McGraw-Hill, 2007.

Investigations in Number, Data, and Space Grade 4. Investigations in Number, Data, and Space Series. Glenview, Ill.: Pearson Education, 2008.

Lesson Activity Book Grade 4. Think Math! Series. Orlando, Fla.: Harcourt School Publishers, 2008.

Math Expressions Grade 4. Math Expressions Series. Boston: Houghton Mifflin, 2006.

Math Grade 4. Math Series. New York: Macmillan/McGraw-Hill, 2005.

Math Masters Grade 4. Everyday Mathematics Series. Chicago: Wright Group/McGraw-Hill, 2007.

Math Trailblazers Grade 4. Math Trailblazers Series. Dubuque, Iowa: Kendall Hunt, 2004.

National Council of Supervisors of Mathematics (NCSM). *Essential Mathematics for the Twenty-first Century: The Position of the National Council of Supervisors of Mathematics.* Lakewood, Colo.: NCSM, 1988.

National Council of Teachers of Mathematics (NCTM). *Curriculum and Evaluation Standards for School Mathematics.* Reston, Va.: NCTM, 1989.

―――. *Principles and Standards for School Mathematics.* Reston, Va.: NCTM, 2000.

―――. *The Role of Technology in the Teaching and Learning of Mathematics.* Reston, Va.: NCTM, 2008.

National Research Council. *Reshaping School Mathematics: A Philosophy and Framework for Curriculum.* Washington, D.C.: National Academy Press, 1990.

Student Guide Grade 4. Math Trailblazers Series. Dubuque, Iowa: Kendall Hunt, 2004.

Teachers Reference Manual. Everyday Mathematics Series. Chicago: Wright Group/McGraw-Hill, 2007.

Think Math! Grade 4. Think Math! Series. Orlando, Fla.: Harcourt School Publishers, 2008.

Usiskin, Zalman. "Groping and Hoping for a Consensus on Calculator Use." *Mathematics Education Dialogues* 2 (May/June 1999): 1.

Chapter 12

Technology and Mathematics in the Middle Grades

Richard Hollenbeck
James Fey

THERE is never a dull moment in teaching mathematics to students in the middle grades. Adolescents seem to arrive at school each day with surprising questions and ideas in their developing minds. But the surprises and changes in mathematics teaching are not limited to the students who enter our classrooms.

Over the past several decades, the emergence of electronic tools has transformed the ways that we can engage students in exploring mathematical ideas and solving mathematical problems. Calculators, notebook computers, and cell phones provide instant access to powerful options for numeric, graphic, and symbolic calculation and to resources on the Internet. Those same tools allow students to communicate ideas and questions to teachers and classmates around the world at all hours of the day and night.

When electronic information technologies are applied to the tasks of teaching, they provide intriguing opportunities for transforming the mathematics learning experience. From computer tutors, virtual manipulatives, and SmartBoards to e-books, simulation applets, and computer-adaptive testing, we have access to teaching tools that were hard to imagine in the chalk-and-talk era.

Many middle-grades mathematics classrooms already provide students with an impressive array of technological tools. In some schools, access to tools is the easy part. Figuring out how to use the tools effectively and appropriately is a far greater challenge. If you and your students had full use of existing mathematical and communication tools, how would such tools change—

- the way that you teach mathematics in the middle grades?
- the way that you assess students' learning?
- the content of your curriculum?

In this chapter, we examine the questions raised by the emergence of technology-rich mathematics classrooms. Our objective is to stimulate thinking and experimentation by individual teachers, mathematics departments, teacher educators, curriculum and test

---

Adapted from Hollenbeck, Richard, and James Fey. "Technology and Mathematics in the Middle Grades." *Mathematics Teaching in the Middle School* 14 (March 2009): 430–35.

developers, researchers, and educational policymakers about the need and direction for change in middle-grades mathematics.

Technology and Mathematics Teaching

The mathematical content of the middle-grades curriculum is drawn from the Geometry, Measurement, Data Analysis and Probability, Number and Operations, and Algebra strands. Currently, there are attractive tools for developing key ideas in each of these topic areas while concurrently supporting students' development of the mathematical practice of using appropriate tools strategically, as elaborated in the Common Core State Standards for Mathematics (Common Core State Standards Initiative 2010, p. 7).

Geometry and Measurement

This is your objective for one day in a middle-grades mathematics classroom: to develop student understanding of the geometric principle that the area of any triangle can be calculated using the formula $A = {}^1/_2 bh$. With that goal in mind, you might ask your students to find the largest triangle that can be drawn inside a given rectangle. In time, students will work through a variety of approaches for solving the problem. They may construct physical models, draw pictures, or present an analytical method for making sense of the task.

Your challenge is to find ways for students to share and explain their solution strategies. Some teachers have discovered that a document camera, a special video camera designed to display printed and handwritten pages and three-dimensional objects, is an effective tool. A document camera allows students to take turns showing calculations and diagrams that support their reasoning. Different pieces of student work can be displayed to compare ideas. You can also take pictures of student work to archive for future reference.

After discussing their initial ideas, you can direct students to a variety of computer applets for additional exploration or reinforcement of their conjectures. For example, Utah State University's National Library of Virtual Manipulatives (NLVM, http://nlvm.usu.edu) contains a virtual geoboard that students can use to generate many examples of triangles that are enclosed within a given rectangle (see fig. 12.1.).

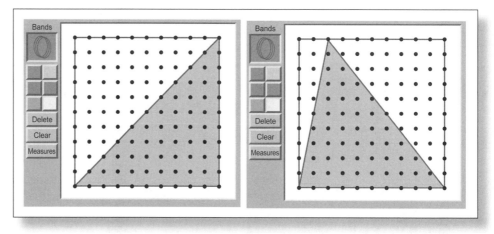

Fig. 12.1. What is the area of the largest triangle that can be enclosed in the given rectangle?

This exploratory work will probably reveal or confirm the principle that for each rectangular configuration, the area of the largest enclosed triangle is equal to half the area of the original rectangle. Students can then explore the same question with a different applet available on the NCTM's Illuminations website (see fig. 12.2). This applet allows students to move one vertex of the triangle without changing the height. They will quickly notice that the shape of the triangle changes but that the base, height, and area do not.

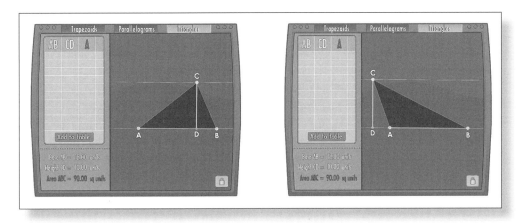

Fig 12.2. How is the area of a triangle related to the length of its base and its height?

By combining visual images and numerical area calculations with some analytic reasoning about the case of right triangles in a rectangle, students are likely to develop a solid understanding of this area formula. At least for triangles with one side along the length or width of the rectangle, they will find the area of the largest inscribed triangle.

Data Analysis and Probability

Data analysis is an important component of the middle-grades mathematics curriculum (NCTM 2000). When students formulate questions and design research plans, statistical software and interactive whiteboards can be used to collect, display, and analyze the data in new and exciting ways.

For example, a middle-grades mathematics class might be interested in exploring how a cell phone conversation adversely affects response time during a concurrent activity. Several online tests of reaction time and gaming systems such as Nintendo's Wii measure response times with millisecond precision. A variety of statistical software packages give teachers and students the power to analyze the resulting data. Typical data analysis programs generate the summary statistics of mean, median, and range. Generating box plots will allow them to compare data sets visually. Students can also investigate how sensitive the mean is to outliers; understanding this principle will help students when choosing appropriate measures of central tendency for data. If the computer software is combined with an interactive whiteboard, the possibility of learning while holding students' attention can be greatly increased. Replacing traditional chalkboards or whiteboards with SmartBoards allows teachers and students to control software programs both from a computer and from the front of the classroom. Thus, the board becomes an interactive focal point of experimentation and discourse about data analysis concepts. This interactive technology can also be applied to classroom investigations in geometry and algebra.

Number and Operations

Proportional reasoning is a core subject in the Number and Operations Standards for middle-grades mathematics—the capstone of the elementary school curriculum and the cornerstone of high school mathematics and science (Post, Behr, and Lesh 1988). The importance of ratios and proportions is enhanced by their use in reasoning about similarity of geometric shapes. This visual representation of proportion in, for example, digital photography and computer graphics provides engaging contexts for student exploration. Imagine a computer display picturing a student standing next to a taller object, such as a climbing wall, and challenging the class to find the height of the wall. Students will probably have intuitive ideas about ways to use the known height of the student to calculate the height of the climbing wall. Then virtual rulers can be used to measure object lengths in pixels, centimeters, or inches. Using several different units of measurement for each object will reveal the invariance of the *ratio* of the two heights. Photo-editing software allows for easy enlargement or reduction of a picture. If the heights of the two objects are measured after each size-change operation, a scatterplot of the measurement pairs will reveal a linear pattern. A spreadsheet, graphing calculator, or computer line-of-best-fit analysis will show how to model that pattern with a linear function of the form $y = mx$. The proportionality relationship expressed in an algebraic form can be used to answer the original question about the height of the climbing wall in a new way.

Algebra

A goal of the middle school mathematics curriculum is to develop students' skills in, as well as an understanding of, solving linear equations. Computer simulations and calculating tools can be used to provide insight into the concepts and skills involved in that process.

For example, one of the most effective ways of thinking about equations and inequalities is the analogy presented by a simple pan balance. A live demonstration with an actual pan balance might be the best way to start, but a computer simulation can also lead students to discover the operating principles that produce equivalent but simpler equations. Given a virtual pan balance provided by an applet (http://nlvm.usu.edu), students can move unknown numeric "weights" to see which moves retain balance but lead to a picture revealing the value of x (see fig. 12.3).

When students understand the basic concept of solving equations and have developed an informal strategy, you can then discuss more efficient solution methods. In most computer algebra systems (CAS), once an equation is entered, it is easy to perform an operation on both sides of the equation. The CAS accurately executes the operations asked of it, often showing a result that is different from what students expected. For example, when beginning students are asked to solve for x in $3x + 5 = 17$, they often try to incorrectly "divide by 3" or "subtract $2x$." The CAS will show the unhelpful results of those moves.

Technology and Mathematics in the Middle Grades

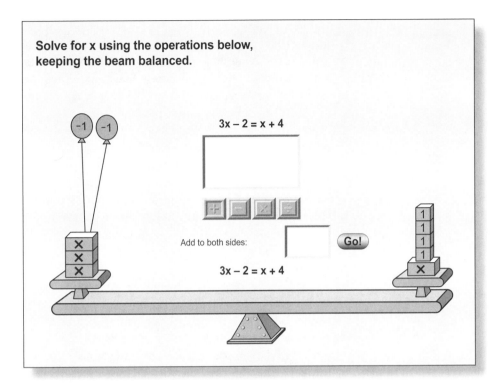

Fig. 12.3. What moves will keep the beam balanced or reveal the value of *x*?

When dividing by 3:

$$\frac{3x+5=17}{3} \rightarrow \frac{3x+5}{3} = \frac{17}{3}$$

When subtracting 2*x*:

$$\begin{array}{r} 3x+5=17 \\ -2x \qquad -2x \\ \hline x+5 = 17-2x. \end{array}$$

Research has shown that when students explore solutions and receive feedback such as that provided by CAS, they quickly develop the understandings and strategic skills that are the desirable goals of instruction.

Technology and the Assessment of Mathematics Knowledge

In the traditional mathematics classroom, assessing student learning was limited largely to quizzes and tests that provided summative descriptions of knowledge. These summative assessments were used primarily to assign course grades. But NCTM's *Principles and Standards for School Mathematics* states, "Assessment should support the learning of important mathematics and furnish useful information to both teachers and students" (NCTM 2000, p. 11). Middle-grades mathematics teachers now have access to innovative technologies that enable them to increase and improve their repertoire of assessment strategies.

For example, several computer-based intelligent tutoring systems provide mathematics assessment and instruction that can be customized for students. These systems track individual student progress and provide additional instruction in areas of need. The assessment and tutoring systems then communicate diagnostic information to teachers about the performance of individual students or an entire class.

Another intriguing assessment tool is a personal response system. At any point in a lesson, students can be asked to use a remote control "clicker" to respond to a teacher's multiple-choice question. Software on the teacher's computer immediately processes student responses and produces summary statistics and charts showing the distribution of student choices among the answer options. Adolescents are often reluctant and uncomfortable when being asked to publicly present their mathematical work. The use of clickers can increase student participation and engagement by providing an anonymous way to assemble student responses to questions. If the results of a question indicate that many students are confused about a topic, teachers can immediately adjust instruction to address the problem.

A different form of personal response technology is provided by devices that allow students to share results of their graphing calculator and computer work. For instance, by scanning submissions of graphs produced to solve an algebra problem, teachers can quickly assess the understandings of many students. Teachers can then select interesting examples of student work to display and analyze in whole-class discussions. By sharing work in this way, teachers gain insight into the knowledge of individual students, and they create powerful opportunities for students to do self-assessment by comparing their own thinking with that of others.

Internet technology has opened up other opportunities for teachers to assess student understanding and skill. For example, blogs allow a unique assessment opportunity in that users can interact with one another by posting comments and questions about a theme. At the start of a new unit, the teacher could ask each student to respond to a question by clicking on the comment link of a blog. Then at different points during the unit, students can revisit the blog to rethink their answer to the same question. By scrolling down a page of comments, the teacher can quickly assess the change over time in students' thinking about a problem.

Many teachers are already using Internet communication to respond to student questions outside of class and collect and respond to electronic submission of homework assignments. As textbooks become increasingly available in electronic (even editable) form, one can imagine this sort of electronic submission of class and homework becoming the norm, just as it is becoming common at the collegiate level.

Technology and the Mathematics Curriculum

The software and graphics capabilities of calculators and computers are particularly well suited to the logical and algorithmic operations of numeric, graphic, and symbolic calculation essential in mathematical work. Numeric functionality performs exact and approximate arithmetic on whole numbers, fractions, and decimals, as well as irrational and complex numbers. Graphical displays help with analysis of data and functions. These tools also display, measure, and transform geometric figures that satisfy prescribed conditions. CAS help solve equations, transform expressions, and test conjectured identities.

With all these possibilities for technology, these questions come to mind:

- What does this current and emerging access to tools for mathematical work imply about our content goals in mathematics teaching?
- Is it still important for all middle-grades students to become proficient in the standard computational algorithms of arithmetic?
- Is it still important for students to become proficient in the routine algebraic operations on expressions, equations, and inequalities?
- How is statistical analysis transformed by access to sophisticated data analysis tools?

Arithmetic in the Future

In mathematics classrooms of the precalculator era, a large portion of instructional time was devoted to training all students in procedures for addition, subtraction, multiplication, and division of whole numbers, common fractions, and decimals and calculations with proportions and percents. Much of the responsibility for developing those skills—especially work involving fractions, decimals, and percents—lay in the middle grades.

That said, if students have access to a calculator, the benefits of knowing the calculation algorithms come into question:

- Is the important objective to develop students' estimation strategies and their disposition to question reasonableness of calculator results? Testing reasonableness of arithmetic calculations seldom involves replicating standard algorithms in one's head or with pencil and paper.
- How does proficiency with standard algorithms contribute to the essential problem-solving skill of deciding which arithmetic operations will yield solutions or at least useful information about the problem?

These are not new questions in mathematics education or in the public discourse about technology and mathematics curricular goals. But the infusion of calculation tools in all aspects of contemporary life makes reconsidering educational objectives a timely discussion.

Algebra in the Future

The case for developing students' proficiency with arithmetic operations and standard algorithms is often justified by the argument that those skills are essential for success in learning algebra. If one thinks about algebra as a collection of syntactic rules for transforming expressions, equations, and inequalities into equivalent forms—unaided by tools such as spreadsheets, CAS, and graphing utilities—the importance of skill in generalized arithmetic procedures is obvious. However, once again, almost anyone who needs to operate on algebraic expressions, equations, and inequalities in technical work will have access to tools that make those tasks routine.

The use of graphing calculators to produce tables and graphs for solving equations and inequalities is widely known and applied. For an example, consider this algebraic problem:

> An amusement park charges $19.95 for individual admissions but offers a group rate of $95.00 plus $13.95 per group member. If a school class is planning an outing to the park, which pricing option is the better choice?

Inspecting tables of values or graphs for the functions $I(n) = \$19.95n$ and $G(n) = \$95.00 + \$13.95n$ shows that the individual price option is the better choice for groups of fifteen or fewer members and that the group option is better for sixteen or more. (See the graph and table in fig. 12.4.)

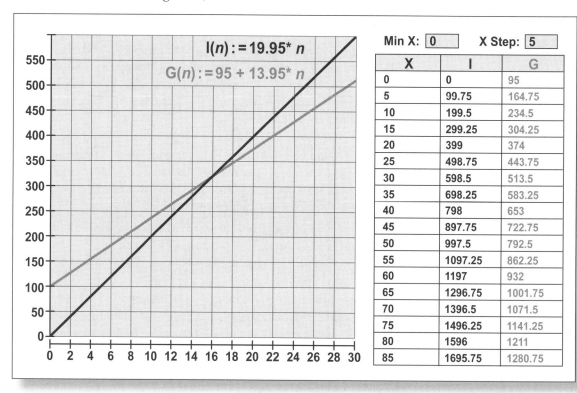

Fig. 12.4. How can you use the tables and graphs to determine the better options for admission to the amusement park?

The break-even point for the two pricing options can also be calculated by solving the equation $\$19.95n = \$95.00 + \$13.95n$. The precise solution can be obtained with CAS by asking it to solve the equation $19.95n = 95.00 + 13.95n$. A CAS can do much of the standard algorithmic calculation that consumes instructional time in traditional algebra courses.

The availability of computer algebra tools suggests that we need to question the goals of middle-grades algebra instruction:

- Does focusing instruction on manipulation of symbolic expressions, equations, and inequalities provide students with the most useful algebraic understanding and skill?
- Is there a productive connection between learning the algebraic skills of manipulating expressions and developing the ability to identify and represent problem conditions in the algebraic forms to which spreadsheet, graphic, and computer algebra system software can be applied?

Data Analysis and Probability in the Future

In much the same way that calculators and computers raise doubts about curricula that focus on procedural skills in arithmetic and algebra, tools that perform calculations in data analysis and probability suggest rethinking the goals of those important strands in middle-grades mathematics. Widely available statistical software allows students to enter data from many different interesting sources or from their own experiments, calculate summary statistics, and display the data with graphics such as line plots, histograms, box plots, and scatterplots. For example, the screen shot in figure 12.5 shows how a scatterplot can be used to look for relationships between nutritional attributes of common fast foods.

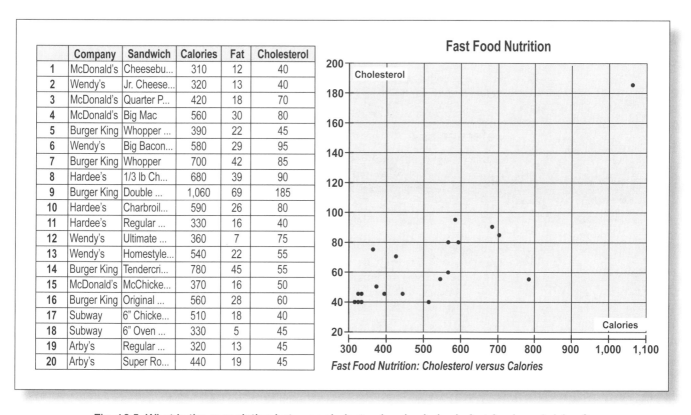

Fig. 12.5. What is the association between cholesterol and calories in fast food sandwiches?

Probability software can help students to simulate experiments with random processes and to do the combinatorial calculations implied by theoretical analysis of those situations. Tools for data analysis and probability calculations are now available to and used by nearly everyone who needs them for problem solving and decision making. This situation raises questions about middle school mathematics:

- How much time should be devoted to developing student skill in statistical calculation and graphing, and how much time should be given to interpreting results of those procedures?
- What experiences with paper-and-pencil calculations and graphing are essential as a foundation for understanding basic concepts and the thoughtful use of statistical tools?
- What is the optimal mix of hands-on experiments and simulations in learning basic probability concepts?

Conclusion

The calculation and computation tools used by the workforce have been adapted to teach mathematics in the middle grades. Appropriate use of these tools and revising curriculum priorities to reflect how mathematical work is done in a technological environment will require extensive and thoughtful study and experimentation. Given the urgency of providing strong mathematical preparation for students who will enter and live in a technologically sophisticated society and workplace, such study and experimentation by all involved in mathematics teaching should be a high priority.

References

Common Core State Standards Initiative. *Common Core State Standards for Mathematics.* Washington, D.C.: National Governors Association Center for Best Practices and the Council of Chief State School Officers, 2010. http://www.corestandards.org.

National Council of Teachers of Mathematics (NCTM). "Illuminations." http://illuminations.nctm.org/ActivitySearch.aspx.

———. *Principles and Standards for School Mathematics.* Reston, Va.: NCTM, 2000.

Post, Thomas R., Merlyn J. Behr, and Richard Lesh. "Proportionality and the Development of Prealgebra Understandings." In *The Idea of Algebra K–12*, 1988 Yearbook of the National Council of Teachers of Mathematics (NCTM), edited by Arthur F. Coxford, pp. 78–90. Reston, Va.: NCTM, 1988.

Utah State University. "National Library of Virtual Manipulatives." http://nlvm.usu.edu.

Chapter 13

Using Computer Algebra Systems to Develop Big Ideas in Mathematics with Connections to the Common Core State Standards for Mathematics

Rose Mary Zbiek
M. Kathleen Heid

TWENTY years ago, talking about high school classrooms whose electronic technology included graphing calculators was quite provocative. Fifteen years ago, a dynamic geometry environment, sometimes coupled with graphing calculators, was state-of-the-art technology. Today, computer algebra systems (CAS) are becoming essential tools for teaching and learning high school mathematics; their trajectory, we hope, will mirror the best of what has happened with calculators in grades K–5 (Chval and Hicks 2012). In this chapter, we illustrate several useful features of CAS, particularly when a CAS is incorporated as part of a classroom tool set. We also address several ways in which CAS can focus our taught and learned curriculum on big mathematical ideas that previously might not have received appropriate attention. Our goal is to stimulate further thinking and experimentation by individual teachers and mathematics departments about the role of CAS in high schools, particularly now in light of the Common Core State Standards for Mathematics (CCSSM; Common Core State Standards Initiative 2010).

CAS is a multirepresentational environment with symbolic, graphical, and numeric capabilities. It links representations so that when users make changes in one representation, other representations automatically update to reflect those changes. For example, using a CAS, a student might enter $f_1(x) = 3x^2 - 1$ symbolically, generate a graph (see fig. 13.1a), and then drag the graph to create a graph that appears wider (see fig. 13.1b) or lower and to the right (see fig. 13.1c). A CAS with multiple linked representations automatically updates the symbolic expression and table, as shown.

Some tools allow for "sliders"—elements that one can add to a file so that a value quickly changes when a user drags or clicks some part of the slider. Figure 13.2 shows the results of using sliders to change the value of a as the amplitude of a sine function

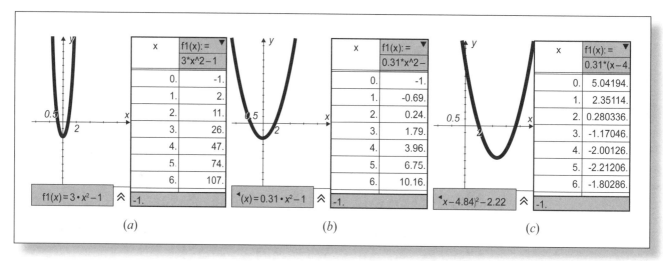

Fig. 13.1. A quadratic function is entered symbolically, and the graphs and tables are produced (a). When the graph is dragged, the symbolic rules and tables change (b and c).

(see fig. 13.2a) and to change the value of L as length to generate rectangles with area 20 square units (see fig. 13.2b). When students combine CAS with other tools and dynamic capabilities, using symbols in powerful ways to capture relationships and predict outcomes becomes an important part of what they learn and do. Hollenbeck and Fey (2009) describe the use of applets and additional activities with dynamic tools. The technological environment that we should seek is a blend of CAS and other technological tools for learning and doing mathematics, similar to the collections described by Hollenbeck and Fey (2012) and advocated by the CCSSM.

Extending the observation in the CCSSM that spreadsheets and CAS can be used

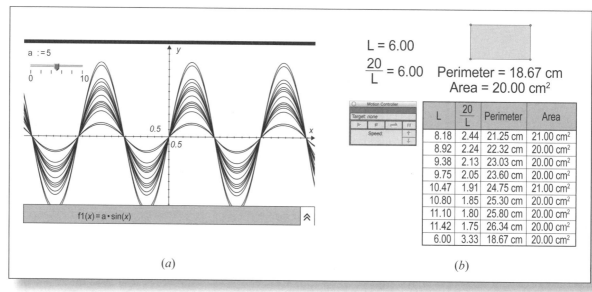

Fig. 13.2. Results of using a slider in a graphing environment to show the effects of changing amplitude (a) and in a geometry environment to generate data for rectangles of area 20 square units (b)

to experiment with symbolic expressions in an effort to understand the behavior of algebraic manipulations, and building on a task used by Kieran and Saldanha (2008), for example, we can use a CAS-capable spreadsheet to create a chart like that shown in figure 13.3. The first column contains a polynomial of the form $x^N - 1$, N a natural number. The second and third columns contain $x - 1$ and the quotient when the polynomial in the first column is divided by $x - 1$. Similarly, the fourth and fifth columns contain $x + 1$ and the result of dividing the first-column polynomial by $x + 1$. This chart is an example of how a CAS-capable spreadsheet can display symbolic expressions and produce the results of symbolic manipulations, which we can use as data in mathematical explorations. The CAS-capable spreadsheet allows students to focus on finding patterns in symbolic forms without the tedium and potential distraction of copying CAS results into a chart. Using the combined symbolic features of a CAS and the visual organization of a spreadsheet places emphasis directly on the CCSSM mathematical practice of looking for and expressing regularity in repeated reasoning.

	B	C	D	E	F
			=b[]/(c[])		=b[]/(e[])
1	x^1 – 1	x – 1	1	x + 1	(x – 1)/(x + 1)
2	x^2 – 1	x – 1	x + 1	x + 1	x – 1
3	x^3 – 1	x – 1	x^2 + x + 1	x + 1	(x^3 – 1)/(x + 1)
4	x^4 – 1	x – 1	x^3 + x^2 + x + 1	x + 1	x^3 – x^2 + x – 1
5	x^5 – 1	x – 1	x^4 + x^3 + x^2 + x + 1	x + 1	(x^5 – 1)/(x + 1)
6	x^6 – 1	x – 1	x^5 + x^4 + x^3 + x^2 + x + 1	x + 1	x^5 – x^4 + x^3 – x^2 + x – 1
7	x^7 – 1	x – 1	x^6 + x^5 + x^4 + x^3 + x^2…	x + 1	(x^7 – 1)/(x + 1)
8	x^8 – 1	x – 1	x^7 + x^6 + x^5 + x^4 + x^3…	x + 1	x^7 – x^6 + x^5 – x^4 + x^3 – x^2…
9	x^9 – 1	x – 1	x^8 + x^7 + x^6 + x^5 + x^4…	x + 1	(x^9 – 1)/(x + 1)
10	x^10 – 1	x – 1	x^9 + x^8 + x^7 + x^6 + x^5…	x + 1	x^9 – x^8 + x^7 – x^6 + x^5 – x^4…

Fig. 13.3. Organized symbolic data produced by a CAS-capable spreadsheet

The kind of CAS-inclusive technological environment possible today allows teachers to emphasize big ideas—ideas different from those considered important when today's teachers (and their teachers) were high school students. To exemplify the difference that CAS and related tools can make in high school mathematics, we describe several ways in which this technology affects which big ideas students can and should learn about algorithm and function.

Using Algorithms

In a non-CAS world, students typically spent hours executing the same procedure for what seemed to them to be quite different tasks—ones that, from a teacher's perspective, required executing the same algorithm. Teachers can use CAS to focus students on either a more targeted or a more global view of their work with symbolic representations, depending on the instructional goal.

Solving Equations

High school algebra historically has centered on solving a few specific types of equations. With CAS, solving equations is a global process of reasoning rather than solely a local process of executing particular steps. This attention to emphasizing and explaining the reasoning underpinning equation solving is central to the Algebra standard of the CCSSM. Suppose that a class is ready to study radical equations. The teacher wants to acquaint students with the conventional technique for solving equations containing one simple radical term; the approach will build on what the students have already learned about linear and quadratic equations and focus their attention on the new part of the radical-equation solution strategy. The CAS facilitates this approach, allowing the teacher to highlight details needed to solve a radical equation and to treat the solution of quadratic equations as an established macroprocedure—in this instance, a global procedure consisting of smaller steps to accomplish a given purpose (Heid 2003). Moreover, the CAS allows the teacher to focus on orchestrating a sequence of old and new steps rather than on executing those steps.

Consider the CAS-based strategy for solving

$$\sqrt{x-5} = x - 7$$

for x over the real numbers shown in figure 13.4. The first three steps of the solution include entering the equation into the CAS, squaring both sides of the equation, and then expanding the results to produce a quadratic equation whose roots include the roots of the original radical equation. The fourth step executes the procedure for solving a quadratic equation, and the fifth and sixth steps check the possible solutions. Using this CAS-based approach with students is important to the teacher who wishes students to understand the following big ideas:

- An initial goal in solving a radical equation is to produce an equation you already know how to solve and whose solution set contains the solutions of the radical equation.
- Particular procedures for solving equations may generate extraneous solutions.

$\sqrt{x-5} = x-7$	$\sqrt{x-5} = x-7$
$(\sqrt{x-5} = x-7)^2$	$x-5 = (x-7)^2$
expand $(x-5 = (x-7)^2)$	$x-5 = x^2 - 14 \cdot x + 49$
solve $(x-5 = x^2 - 14 \cdot x + 49, x)$	$x=6$ or $x=9$
$\sqrt{x-5} = x-7 \mid x=6$	false
$\sqrt{x-5} = x-7 \mid x=9$	true

Fig. 13.4. CAS-based solution of a radical equation

Using Computer Algebra Systems to Develop Big Ideas in Mathematics

In the process of generating this CAS-assisted solution, students' attention and energy can be focused on deciding on the steps of the solution, instead of on executing those steps, and on recognizing that squaring both sides of an equation can result in extraneous roots. In this way, the CAS can help students integrate old and new aspects of the process into a new global view of solving equations.

A CAS also allows students to see the generality of equation-solving strategies. Figure 13.5 uses a CAS-capable spreadsheet and the general solution strategy illustrated in figure 13.4 to solve the equations

$$\sqrt{x-5} = x-7, x-2=0, x^2-4=0, x^3=8, \text{ and } \sqrt{x-5} = 4-\sqrt{x}.$$

The equation to be solved appears in the first column, the equation resulting from squaring both sides appears in the second column, its expanded form appears in the third column, and the potential solution appears in the fourth column. The first five rows in the spreadsheet show results obtained from solving over the real numbers (using a SOLVE command), whereas the sixth row shows results obtained from solving over the complex numbers (using CSOLVE). The point is not that students should use this technique to solve all equations, but that they can view procedures as generalizable and do not need to tailor them precisely to each different problem type. The radical-equation procedure applies to a large range of equations, even though some of the equations contain no radical expression. The generality of this equation-solving technique raises an opportunity for a class discussion of the appropriateness and efficiency of using this technique for various types of equations. Using a CAS-capable spreadsheet in this way, students can test the generality of other symbolic strategies in ways that help them see algebra as a systematic, connected area of mathematics.

	A	B	C	D	E
◆		=a[]^2	=expand(b[])		
1	√(x−5)=x−5	x−5=(x−7)^2	x−5=x^2−14*x+49	x=6 or x=9	
2	x−2=0	(x−2)^2=0	x^2−4*x+4=0	x=2	
3	x^2−4=0	(x^2−4)^2=0	x^4−8*x^2+16=0	x=−2 or x=2	
4	x^3=8	x^6=64	x^6=64	x=−2 or x=2	
5	√(x−5)=4−√(x)	x−5=(√(x)−4)^2	x−5=x−8*√(x)+16	x=441/64	
6	x^3=8	x^6=64	x^6=64	x=1−√(3)*i or x=1+√(3)*i...	
7			x=1−√(3)*i or x=1+√(3)*i or x=−1−√(3)*i or x=−1+√(3)*i or x=−2 or x=2		

Fig. 13.5. CAS-enabled spreadsheet showing results of applying the "radical equation solution algorithm" to different types of equations

What are the advantages of using the symbolic calculator to develop students' understanding of how to solve tasks like this? It accentuates the algorithm instead of the calculation. With the calculator as the executor, the steps in the solution are brought to the fore. Students can see solving radical equations as a macroprocedure, requiring the microprocedures of isolating the radical term, squaring both sides of the resulting equation, expanding expressions as needed, and solving equations of a familiar form. Students can internalize the global process without being caught up in the details. Moreover, with reflection on their results, students can see this procedure as a slight variation on the methods of solution they had mastered for linear and quadratic equations.

Solving Systems

Although using the CAS as a symbolic manipulation assistant can focus students' attention on the algorithm, it often requires a greater capability in symbolic and graphical reasoning. For example, consider generating the solution of the following system of inequalities: $x^2 + y^2 \leq 4$ and $y \geq x^2$. For this task, students might think a coordinate geometry solution is in order. The ordered pairs (x, y) that fit the constraints represent those points that are on or in the interior of the circle with center at $(0, 0)$ and radius 2 and that are on or above the parabola given by $y = x^2$ (see fig. 13.6.).

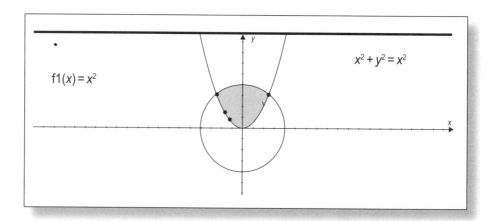

Fig. 13.6. Graph of the solution to the system of inequalities $x^2 + y^2 \leq 4$ and $y \geq x^2$

Although some students may have thought of the graphical approach, they might have found it challenging or tedious to develop the graphs and identify the region representing the solution, or they might have the impression that a symbolic approach is always favored. In fact, a symbolic approach to solving this system of inequalities with most CAS requires only one command. Figure 13.7 shows the output when one CAS was issued a command to solve this system. This symbolic representation of the solution of this system is more complicated than the graphical image because of the need to express the solution in terms of simultaneous constraints on x and y. In this instance, six inequalities are required. The problem and its solutions bring home the value of graphical representation in its capacity to account for two conditions at once.

By generating and comparing solution strategies, through the use of multirepresentational tools, students develop a big idea about solution strategies:

- Graphical representations are powerful because they account for several constraints at once.

```
solve (x²+y²≤4 and y≥x²,{x,y})
    x−√(4−y²)≤0 and 0≤x≤√y and 0≤y≤2 or x+√(4−y²)≥0 and −√y≤y<0 and 0≤y≤2)
```

Fig. 13.7. CAS output for solving the system $x^2 + y^2 \leq 4$ and $y \geq x^2$

This is a particularly important understanding for students who believe that they are to use symbolic methods with particular types of tasks, that other methods are not allowable, and that they must memorize which technique to use with which type of task.

Using the CAS helps us understand better the promise and limitations of different representations and strategies. When used in ways that emphasize big ideas about algorithms, the CAS helps students advance into the world of metamathematics—not just by doing symbolic manipulation but also by thinking about the nature of symbolic procedures.

Expanding the Concept of Function

A CAS-inclusive environment is conducive to expanding the idea of function beyond the learners' typical experiences: finding the domain of a function and composing functions. As one student accurately observed, finding the domain was one of the few lessons in her algebra experience that *needed* functions with rules such as these:

$$f(x) = \frac{x+2}{x^3+8}, \; g(x) = \sqrt{7-2x}, \text{ and } h(x) = \frac{x}{\sqrt{x^2+1}}.$$

When students did not need to be concerned about zero denominators and negative radicands, finding the domain was not mathematically interesting or challenging. Composing functions when tasks include only linear and quadratic functions allows for reasonably easy by-hand symbolic manipulation but lets students ignore the mathematical importance of domains and ranges. By capitalizing on CAS multiple representations and looking broadly at function, students encounter two newly important big ideas, one related to function composition and one related to geometric transformations.

Function Composition

Using CAS also helps to shift students' attention from the repetitive and tedious manipulation acts that constitute many lessons about finding the domain and composing functions to a focus on a big idea that unites the two topics:

- Function composition is an operation on functions that creates a function from two given functions under the appropriate conditions on the domains and ranges of the given functions.

Consider generating expressions for composition of two simple functions:

$$f(x) = x^2 \text{ and } g(x) = \sqrt{x},$$

where x is a real number. Entering $f(g(x))$ yields x (see the third line of fig. 13.8). At this point, some CAS give a warning, such as "Domain of the result might be larger than the domain of the input." Why does this message appear? This question can draw students into thinking about domain restrictions that remain masked when their attention is on how to compute the composition expressions. Similarly, entering $g(f(x))$ yields $|x|$ (see the fourth line of fig. 13.8). Why is this answer not simply x? Implementing composition tasks that involve a wider range of functions and engaging students in discussion of CAS results change the big ideas that students develop about function composition and its relationship to domain in particular and about function in general.

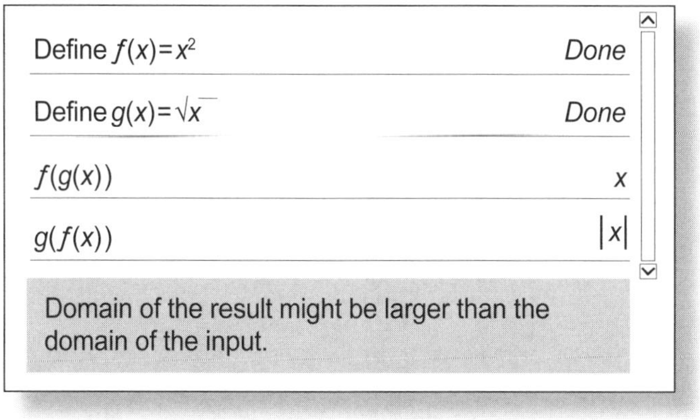

Fig. 13.8. Expressions for compositions of function, and warning message following $f(g(x))$ as produced by CAS

As noted earlier, a CAS with sliders allows students to connect actions on graphical representations with actions on symbolic representations. Figure 13.9 shows four sliders: one each for a, b, c, and d. Clicking on the up or down arrowheads for one of the sliders causes the values of a, b, c, or d to increase or decrease, respectively. For example, given the functions f_1 and f_2 with rules $f_1(x) = x^2$ and $f_2(x) = af_1(b(x - c)) + d$, increasing d causes the graph to move up by the number of units equal to the change in d. Changing d from 1 to 6 moves the graph up five units. Experimenting with the dynamic sketch illustrated in figure 13.9, students typically conjecture about the effects of changing each of the four parameters; for example, changing the values of d and c creates graphs that appear to move up and down or left and right, respectively. Changing the value of a and b leads to graphs that seem to expand or contract vertically or horizontally.

We can take advantage of CAS to use the symbols to reason about these observations. For example, the calculations in figure 13.10a begin by defining the basic function

$f_1(x) = x^2$. The corresponding general function is $f(x) = a(b(x-c)^2) + d$. The constant elements k in $f_2(x)$ and $k+1$ in $f_3(x)$ correspond to dragging the slider to increase the value of parameter d by 1, from k to $k+1$. The last line of figure 13.10a illustrates how increasing the parameter by 1 yields a difference of 1 between $f_2(x)$ and $f_3(x)$, for any value of x. The constant difference of 1 in the last line is the symbolic version of the visual observation that every point on the new graph is one unit above the corresponding point on the old graph.

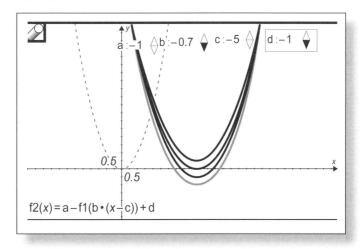

Fig. 13.9. Illustrations of how the graphs appear when a slider is used to change the values of d in $g(x) = af_1(b(x-c)) + d$

In figure 13.10b, adding 1 to the value of a results in the expressions given for $f_4(x) = kf_1(b(x-c)) + d$ and $f_5(x) = (k+1)f_1(b(x-c)), + d$. The difference between $f_4(x)$ and $f_5(x)$, the output values for a value of x, is $b^2x^2 - 2b^2cx + b^2c^2$. This nonconstant difference between corresponding output values when a increases by 1 explains why some points of the new graph are far above the corresponding points on the original graph, while other points of the new graph are close to their corresponding points on the original graph. For example, for the function rules for $f_4(x)$ and $f_5(x)$ when $b = 1$ and $c = 0$, the symbolic results for x-values of 0, 1, and −1 in figure 13.11 show how the vertical differences between the new graph point and the corresponding point on the original graph are different for $x = 0$ and $x = 1$, but the same for $x = 1$ and $x = -1$. Interestingly, the symbolic observation that differences in the output values are the same for $x = 1$ and $x = -1$ is consistent with the symmetry of these graphs with respect to the y-axis. This fine-grained analysis of the symbolic results of increasing the value of a parameter by 1 gives an exactness to the approximate results we have from the graph. If the value of parameter d increases by 1, the graph appears to be translated up approximately one unit. The symbolic results confirm that the translation is by exactly one unit. Similarly, graphical observations of the effect of increasing the value of a by 1 can be corroborated symbolically. Using CAS to analyze the situation graphically and symbolically engages students in the CCSSM mathematical practice of constructing and critiquing arguments.

Curriculum Issues in an Era of Common Core State Standards for Mathematics

Define $f1(x)=x^2$	Done
Define $f2(x)= a \cdot f1(b \cdot (x-c))+k$	Done
Define $f3(x)= a \cdot f1(b \cdot (x-c))+k+1$	Done
$f3(x)-f2(x)$	1

(a)

Define $f1(x)=x^2$	Done
Define $f4(x)= k \cdot f1(b \cdot (x-c))+d$	Done
Define $f5(x)= (k+1) \cdot f1(b \cdot (x-c))+d$	Done
$f5(x)-f4(x)$	$b^2 \cdot x^2 - 2 \cdot b^2 \cdot c \cdot x + b^2 \cdot c^2$

(b)

Fig. 13.10. Symbolic work showing differences between original and new output values when the values of d and a in $f(x) = a(b(x - c)^2) + d$ increase by 1

Define $f1(x)=x^2$	Done
Define $f4(x)= k \cdot f1(b \cdot (x-c))+d$	Done
Define $f5(x)= (k+1) \cdot f1(b \cdot (x-c))+d$	Done
$f5(1)-f4(1)$	$b^2 \cdot (c-1)^2$
$f5(-1)-f4(-1)$	$b^2 \cdot (c+1)^2$
$f5(0)-f4(0)$	$b^2 \cdot c^2$

Fig. 13.11. Symbols illustrating that when $b = 1$ and $c = 0$ adding 1 to the value of a in $f(x) = a(b(x - c)^2) + d$ produces the same differences in the output for $x = 1$ and $x = -1$ but not for $x = 0$

Geometric Transformations

The idea that students can quickly produce the graph of a complicated function if they know how to transform the graph of a related function is something that they should relate to their study of transformations in geometry. CAS-supported work can help students not only appreciate transformations as a type of function but also experience transformations through multiple representations, such as matrix representations, geometric images, and systems of equations. For example, the matrix

$$\begin{bmatrix} -1 & 0 \\ 0 & -1 \end{bmatrix}$$

as well as the system of equations

$$\begin{cases} x' = -x \\ y' = -y \end{cases}$$

and the geometric sketch in figure 13.12 capture a rotation of 180° about the origin. Experience with CAS facilitates use of these different representations—and particular symbolic representations—to answer such questions as which, if any, points are fixed under this rotation. A solution using a one-step command is shown in figure 13.13.

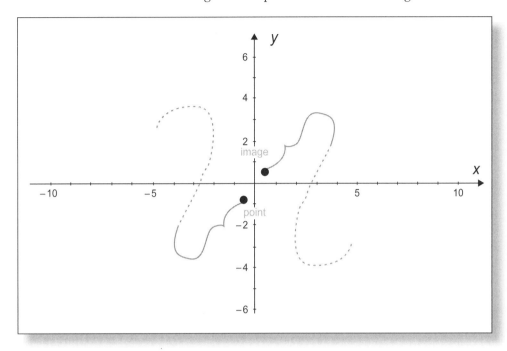

Fig. 13.12. Trace of point and its image under a rotation of 180° about the origin

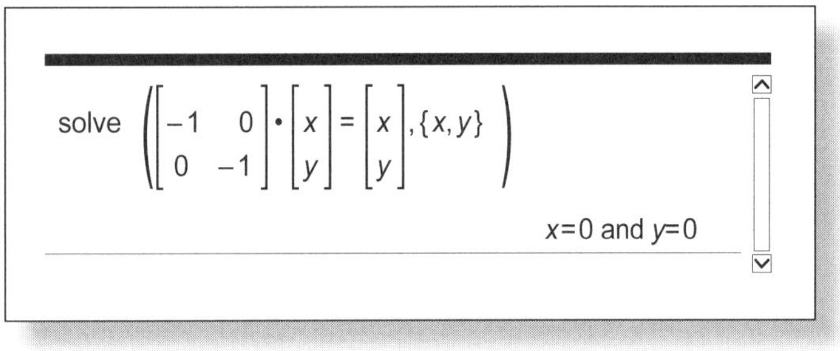

$$\text{solve}\left(\begin{bmatrix} -1 & 0 \\ 0 & -1 \end{bmatrix} \cdot \begin{bmatrix} x \\ y \end{bmatrix} = \begin{bmatrix} x \\ y \end{bmatrix}, \{x, y\}\right)$$

$x=0$ and $y=0$

Fig. 13.13. One-step CAS command to determine that (0, 0) is a fixed point under a rotation of 180° about the origin

The examples regarding function illustrate how students with CAS as part of a mathematics tool set can encounter the related big ideas:

- There are many types of functions, some of which have multiple representations that need not be limited to graphs, symbols, and numbers.
- Properties and reasoning observed in one representation might be explained by reasoning in another representation.

Further examples of what CAS might do and how these features have been used in secondary school mathematics can be found in *Computer Algebra Systems in*

Secondary School Mathematics Education (Fey et al. 2003) and in the November 2002 issue of *Mathematics Teacher* (the theme of this focus issue is computer algebra systems).

Conclusions

In the future, secondary school mathematics programs that make use of CAS and underscore big ideas can alter and sharpen current emphases in school mathematics. As teachers and students work on developing symbolic procedures, they can use CAS to emphasize selected aspects of the procedures by allocating familiar or less important aspects to automated symbolic manipulation. Teachers can direct students' attention to novel aspects of new procedures and to how these new procedures connect to what students already know and can do. CAS will also enable the learning of generalizable procedures in lieu of a collection of specialized procedures. Finally, availability of CAS will sharpen the role of multiple representations in the secondary school mathematics classroom. Use of CAS, its symbolic capacity as well as its linked graphical and numerical capacities, will lead to new expectations of students. They will be expected to move fluidly among representations, easily explain graphical and numerical results symbolically, and reason about how mathematical properties represented symbolically are reflected in graphical and numeric representations. The examples in this chapter illustrate how readily CAS-active tasks provide these experiences while addressing CCSSM content standards and engaging students in CCSSM mathematical practices. Expectations in the CAS-present classroom of the future will revolve around robust and connected conceptual and procedural understanding, making mathematics more accessible, more enjoyable, and very different from the early days of classroom technology.

References

Chval, Kathryn B., and Sarah J. Hicks. "Strategically Using Calculators in the Elementary Grades." In *Curriculum Issues in an Era of Common Core State Standards for Mathematics*, edited by Christian R. Hirsch, Glenda T. Lappan, and Barbara J. Reys, pp. 125–137. Reston, Va.: National Council of Teachers of Mathematics, 2012.

Common Core State Standards Initiative. *Common Core State Standards for Mathematics*. Washington, D.C.: National Governors Association Center for Best Practices and the Council of Chief State School Officers, 2010. http://www.corestandards.org.

"Computer Algebra Systems." Special issue, *Mathematics Teacher* 95 (November 2002).

Fey, James T., Al Cuoco, Carolyn Kieran, Lin McMullin, and Rose Mary Zbiek, eds. *Computer Algebra Sysytems in Secondary School Mathematics Education*. Reston, Va.: National Council of Teachers of Mathematics, 2003.

Heid, M. Kathleen. "Theories That Inform the Use of CAS in the Teaching and Learning of Mathematics." *CAS in Mathematics Education*, edited by James T. Fey, Al Cuoco, Carolyn Kieran, Lin McMullin, and Rose Mary Zbiek, pp. 33–52. Reston, Va.: National Council of Teachers of Mathematics, 2003.

Hollenbeck, Richard, and James Fey. "Technology and Mathematics in the Middle Grades." In *Curriculum Issues in an Era of Common Core State Standards for Mathematics*, edited by Christian R. Hirsch, Glenda T. Lappan, and Barbara J. Reys, pp. 139–148. Reston, Va.: National Council of Teachers of Mathematics, 2012.

Kieran, Carolyn, and Luis Saldanha. "Designing Tasks for the Codevelopment of Conceptual and Technical Knowledge in CAS Activity: An Example from Factoring." *Research on Technology and the Teaching and Learning of Mathematics*, Vol. 2, *Cases and Perspectives*, edited by Glendon W. Blume and Mary Kathleen Heid, pp. 393–414. Charlotte, N.C.: Information Age Publishing, 2008.

Section VI
Learning Progressions in School Mathematics: The Case of Statistics

Introduction

Situations involving statistical information confront youngsters and adults alike, daily. Making sense of data and dealing with uncertainty are skills essential to being a wise consumer, an enlightened citizen, and an effective worker or leader in our data-driven society. The National Council of Teachers of Mathematics (NCTM) signaled the importance of statistics education as an integral part of the mathematics curriculum in its *Curriculum and Evaluation Standards for School Mathematics* (NCTM 1989). Subsequent recommendations in state frameworks, NCTM's *Principles and Standards for School Mathematics* (2000), and most recently the Common Core State Standards for Mathematics (CCSSM; Common Core State Standards Initiative 2010), underscored the importance of statistics and probability in school mathematics programs.

In 2007, the American Statistical Association released the *Guidelines for Assessment and Instruction in Statistics Education (GAISE) Report* (Franklin et al. 2007). That report gives learning progressions for important ideas of statistics organized into three developmental levels—A, B, and C. Although these three levels may parallel the standard grade bands (elementary, middle, and high school), they derive from students' prior statistical experiences rather than from grade level. The three companion chapters in Section VI illustrate how to use the GAISE report to shape a coherent grades K–12 development of basic ideas related to distributions of data-based variables. These chapters give concrete examples, for statistics topics, of learning trajectories similar to those Confrey and Krupa outlined in this volume's introduction.

The chapter by Franklin and Spangler illustrates the statistical problem-solving process: designing the questions to answer, collecting and representing the data, analyzing the data, and interpreting the data in the contexts of categorical and discrete numerical data appropriate for the elementary grades (Level A). They develop the concept of "fair share" as the mean value for a set of numeric data. The chapter by Kader and Mamer extends the ideas in the Franklin and Spangler chapter to appropriate middle school (Level B) statistical problems focusing on the mean as the balance point of a distribution and on methods for measuring variation in data from the mean. The chapter focuses on how to help students become adept at connecting numerical summaries to graphical representations. Finally, the chapter by Scheaffer and Tabor uses statistical problem solving to illustrate teaching about distributions of means and proportions and how one can use these distributions to make sound comparative inferences.

As you read these three chapters, think about their implications for designing coherent and connected learning progressions of important statistical ideas in the CCSSM across your district's K–12 mathematics program.

Questions for Reflection and Collective Discussion

1. Why should data analysis, statistics, and probability have a more prominent place in today's grades K–12 mathematics curriculum?

2. To what extent do all your students have opportunities to become proficient in the statistical problem-solving process? Where, if at all, might you consider strengthening your program?

3. Statistician David Moore described data as "numbers with a context." How does each of the three chapters reveal this simple idea?

4. The authors of the first chapter recommend that we expose elementary school students to both categorical and discrete quantitative data. What appropriate, interesting contexts involving *categorical* data lend themselves to involving elementary school students in the statistical problem-solving process? What interesting, appropriate *discrete quantitative data* contexts lend themselves similarly to students' engagement?

5. Important ideas in statistics are the *shape*, *center*, and *spread* of distributions. Describe how your grades K–5, 6–8, and 9–12 mathematics programs develop these ideas. How, if at all, could you use learning progressions of these ideas to help fill gaps or eliminate unnecessary repetition?

6. Scan the Common Core State Standards for Mathematics (CCSSM) for grades K–8. At what grades does this document treat ideas of data and data analysis?

7. Choose a particular statistical topic you teach, and describe how that topic connects to other mathematics you teach. You may wish to use a *concept map* to show those connections.

8. Look back at the chapter(s) you read in this section, and identify specific mathematical practices that they exemplify. Compare your findings with those of your colleagues.

9. The GAISE document is available at www.amstat.org/education/gaise/. Scan this document's organization and content. How could you use the document as a resource in your teaching? As a resource for grades K–12 curriculum planning in relation to the CCSSM?

10. Select an appropriate statistical idea from your state mathematics standards and content expectations or from the CCSSM. Work with colleagues to design and teach a lesson on that idea, using the statistical problem-solving approach outlined in these articles. Reflect and report on how the experience affected teaching and students' engagement and learning. On the basis of this experience, what ongoing curriculum or instructional changes might you make?

11. How do your state standards for statistics and probability compare with the corresponding CCSSM standards (http://www.corestandards.org)?

12. How does your grades K–12 mathematics program's treatment of statistics and probability align with the Statistics and Probability Standard in the CCSSM?

13. In planning for implementing the CCSSM in middle school and high school, it may be helpful to recall Howard Fehr's (1972, reprinted 2006) seminal article "The Present Year-Long Course in Euclidean Geometry Must Go." Fehr argued that every year of study from seventh grade through high school should include geometry, connected increasingly with algebra. This idea is particularly attractive and feasible in states such as Michigan, where 34% of the current Michigan High School Geometry Standards and Expectations are now found in the CCSSM's middle grades standards (Hirsch and Edson 2011). A possible corollary to Fehr's assertion is that similar year-long courses in algebra should also

go, and that each year's study should focus on algebra with strong connections to both geometry and statistics, thereby capitalizing on time savings in geometry. What would be some advantages of such a curriculum organization, particularly in designing a coherent, connected statistics strand for the high school level? What are possible disadvantages?

References

Common Core State Standards Initiative. *Common Core State Standards for Mathematics.* Washington, D.C.: National Governors Association Center for Best Practices and the Council of Chief State School Officers, 2010. http://www.corestandards.org.

Fehr, Howard F. "The Present Year-Long Course in Euclidean Geometry Must Go." *Mathematics Teacher* 65 (February 1972): 102, 151–54. Reprinted, *Mathematics Teacher* 100 (October 2006): 165–68.

Franklin, Christine, Gary Kader, Denise Mewborn, Jerry Moreno, Roxy Peck, Mike Perry, and Richard Scheaffer. *Guidelines for Assessment and Instruction in Statistics Education (GAISE) Report: A Pre-K–12 Curriculum Framework.* Alexandria, Va.: American Statistical Association, 2007.

Hirsch, Christian R, and Alden J. Edson. "The Shape of Geometry in an Era of Common Core State Standards for Mathematics." Presentation at the Conversations among Colleagues Conference: Common Core State Standards—Implications for Mathematics Education, Grand Valley State University, February 5, 2011. http://www.gvsu.edu/math/events/mathinaction/handouts/2011/Hirsch_Edson_CAC_2011.pdf.

National Council of Teachers of Mathematics (NCTM). *Curriculum and Evaluation Standards for School Mathematics.* Reston, Va.: NCTM, 1989.

———. *Principles and Standards for School Mathematics.* Reston, Va.: NCTM, 2000.

Chapter 14

Statistics in the Elementary Grades: Exploring Distributions of Data

Christine A. Franklin
Denise A. Spangler

SCHOOLTEACHERS have long engaged elementary students in collecting and analyzing data but have often neglected to involve students in formulating the questions to be answered (so that the data are relevant and meaningful to students) and to provide opportunities for students to interpret data they have collected in light of their original question. The *Guidelines for Assessment and Instruction in Statistics Education (GAISE) Report* (Franklin et al. 2007) was an influential resource in the writing of the Common Core State Standards for Mathematics (CCSSM; Common Core State Standards Initiative 2010). Of particular importance for teaching is the GAISE framework that outlines a four-step statistical problem-solving process that should be at the forefront of all data analysis scenarios:

1. Formulate a question that can be addressed with data.
2. Collect data to address the question.
3. Analyze the data.
4. Interpret the results.

Elementary school students (level A as described in the GAISE report) should be exposed to data analysis situations involving both *categorical* and *quantitative* data. Much of the data we collect in elementary schools is categorical. That is, students choose from various *categories* to respond to questions instead of giving numerical answers. For example, most questions such as "What is your favorite … (ice cream flavor, color, school lunch choice, season, etc.)?" elicit categorical responses. Questions that elicit numerical (or quantitative) responses should generally be limited to discrete situations at the K–5 level. For example, questions such as "How many chips are in a typical chocolate chip cookie?" or "How many candies are in a typical single-serving package?" can generally be answered

---

Adapted from Franklin, Christine A., and Denise S. Mewborn. "Statistics in the Elementary Grades: Exploring Distributions of Data." *Teaching Children Mathematics* 15 (August 2008): 10–16.

Categorical Data: The Shoe Problem

by using whole numbers to count. In this chapter, we trace one example using categorical data and another using discrete quantitative data and follow each through the statistical problem-solving process.

1. Formulate Questions.

Most elementary school students are curious about their peers and, in particular, what their peers are wearing. One question that a group of students might ask is "What is the most popular type of shoe in our class today?" This question could have a variety of links to the students' lives. For example, how many students are prepared to go to physical education class without changing shoes? Other students might want to wear slip-on shoes (to avoid tying their shoes), but their parents do not think slip-ons are safe for school activities. Finding out how many students are wearing slip-on shoes provides data to answer the question, "Are slip-on shoes really as popular as we think they are?" A teacher might pose the question, "If you were to advise the local shoe store on the type of shoe the store should aim to have in stock, what would you recommend?"

2. Collect and Represent the Data.

Multiple ways exist to classify shoes and to collect data on the basis of different classifications. One option is to have each student remove one shoe and place it in a pile in the middle of the room. Starting with the pile of shoes allows students to grapple with the question of what the categories might be. To determine whether the categories will capture all shoes in the class, they can look at the range of shoes and the shoes still on their feet. Students may suggest two categories, such as "shoes that tie" and "shoes that do not tie." Discussing what is included in the "shoes that do not tie" category (e.g., buckle, Velcro, slip-ons) will help students understand that the two categories are mutually exclusive (nonoverlapping) and that the categories account for all possibilities. Students may also suggest more elaborate categories according to how the shoes fasten, their color or material, or other features. It is important to have students assess whether the categories cover all shoe types in the class (which can be done by asking students to look at their own shoes and determine into which category they would place them). The essential idea here is that whatever classification is used, the resulting data will *vary*. That is, not all students will be wearing the same type of shoes.

3. Analyze the Data.

Once they have decided on the categories, students can sort the pile of shoes and make a graph on the floor. Taping a grid to the floor will help ensure that shoe size does not affect the height of the bar in each category. Students can then replicate the graph by placing sticky notes on a chart or by coloring boxes on graph paper. Each of these activities provides a representation of the *distribution* of the shoes. In this case, the distribution indicates different shoe categories and the number of shoes within each category. Using various representations (graphical and numerical) for summarizing data distribution is one of the most important concepts in statistics.

4. Interpret the Results.

After creating a data representation, students should focus on the shoe distribution. Elementary teachers often neglect to guide students to look beyond the "pictures" the students have created. The data picture is sometimes hung on the classroom or hallway wall with no discussion of what information students gained from collecting data and answering the originally formulated questions. After summarizing the data, encourage students to consider questions such as the following:

- Do all categories contain about the same number of shoes?
- Do some categories contain more shoes or fewer shoes than others?

At this stage of the problem-solving process, answering the original question should become the students' focus: "Which type of shoe is most popular in our class today, and what would you recommend to the local shoe store?" The answers may differ, depending on which categories were selected. For example, if "tie" and "don't tie" were the categories, shoes that do not tie may be the most popular. But if the categories were "tie," "buckle," "Velcro," and "slip-on," the answer may be shoes that tie.

Although students have now answered the original question, do not let the activity end here. More mathematical exploration can be done with the data. For instance, to solidify students' understanding of their distribution representation, ask them to locate themselves in the representation. If multiple representations have been created, focus on a representation (for example, a bar graph) where they cannot determine precisely which sticky note is theirs. In contrast, with the graph from the shoe data, students can easily identify which data point is theirs. If a category has only one data point in it (such as "shoes that buckle"), students may also be able to identify the owner of the data piece. Identifying an individual data piece is an important beginning point for young children as they transition from seeing individual data points to seeing a distribution as a whole. Now is also the time to push students' thinking with extension questions:

- Would we expect different results if we collected these data from another class at a different grade level in our school?
- How might the results look different? Why? (Is type of shoe related to age? Might we see fewer shoes that tie in kindergarten and more slip-on shoes in sixth grade?)
- What if we collected these data at a school in Hawaii? In Canada's Northwest Territories?
- What if we collected data in January instead of September?
- If we collected the data from workers who are building a new school, would we see differences? Why?

Here, the emphasis is on getting students to note reasons for *differences in distributions of data*, such as weather, grade level, or occupation.

Use this opportunity to encourage students to find other questions to answer from the collected data. For example, students might suggest that they can determine how many more people are wearing tie shoes than buckle shoes, or they can determine how many people in the class are wearing slip-on shoes or shoes that fasten with Velcro. Having students write these kinds of questions prepares them for the questions they will meet on standardized tests and also helps them see the link between the question they have asked, the operations needed to answer the questions, and the data from the graph that must be used to answer the questions.

A note of caution: Teachers often encourage students to find the mean as a numerical summary with categorical data. As the next case demonstrates, the mean requires having numerical values for the variable of interest. With a categorical data set, we have categories (e.g., "tie," "slip-on," "Velcro") rather than numerical values. Finding the mean of the *frequency counts* for the different categories—for instance, computing the mean of the number of shoes in each category—is a typical mistake. This "mean" has no significance to the original question ("What is the most popular type of shoe in our class today?") or to the data and, thus, it is inappropriate to ask children to determine the mean of a set of categorical data. The following example illustrates a proper setting for using the mean as a numerical summary.

Numerical Data: The Soccer Problem

Level A statistical questions involving numerical data allow students to develop an interpretation of the mean and begin to explore quantifying variability in the data. Sports provide a convenient context to analyze numerical data.

1. Formulate Questions.

Soccer is a popular sport for both girls and boys; the number of goals scored by soccer teams on a particular weekend is a reasonable student curiosity that would lead to the question of how many total goals are scored in soccer games. (Note that for the remainder of the chapter "total goals" in a game will be referred to as the *score*.)

2. Collect and Represent the Data.

Involving students in deciding what data will be collected is important. A class discussion might lead to the decision to have all students who play on a soccer team report the score in the game they played over the weekend. Make the students aware that only one person per team reports the score to avoid duplicating data. Students should also discuss why it is a good idea to collect data on this past weekend's games rather than reporting the highest score per game of the season. (The latter could result in overestimating a typical soccer game score.) In this investigation, each game score will be represented with a tower of cubes. Figure 14.1 shows this representation of data for the scores from nine games.

3. Analyze the Data.

Begin by asking students what they notice about the data. Students will ptrobably report the lowest score of a game, the highest score of a game, and the scores that are the same. For ease of comparison, students may suggest arranging the towers from smallest to largest and recording the numerical value for the scores (see fig. 14.2).

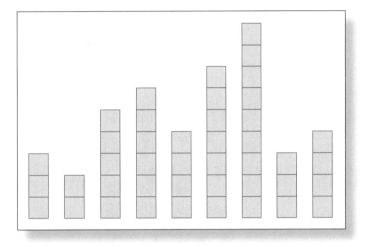

Fig. 14.1. Cube representation for nine soccer games' scores

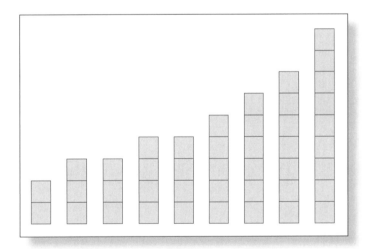

Fig. 14.2. Ordered cube towers

Students should note that the scores are not all the same. That is, scores *vary* from one game to another. In this case, scores vary from 2 to 9. We might ask, "Based on all the goals scored from these nine games, what would be the game score if all games resulted in the same score?" This score is called the *fair*, or *equal*, *share* value for the data. A process for determining this value follows:

- The first step is to combine the scores from all games into one large group of individual goals.
- A total of 43 goals were scored in the nine games (see fig. 14.3).
- Remove nine cubes from the group. These nine cubes represent a single goal scored in each of the nine games. Thirty-four cubes remain in the group (see fig. 14.4).
- Next, remove another nine cubes from the group to represent an additional goal scored in each of the nine games (see fig. 14.5). Continue this process until no cubes are left (see fig. 14.6).

Fig. 14.3. All 43 goals scored

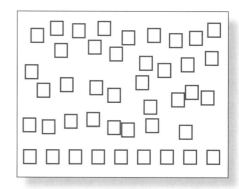

Fig. 14.4. Step 1 of cubes

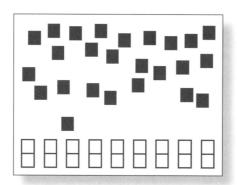

Fig. 14.5. Step 2 of cubes

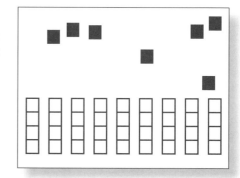

Fig. 14.6. White cubes for each game with seven remainder cubes

Thus, the fair share value for these data is $4^7/_9$. Because this quantity is not a whole number, interpreting its value in the context of this problem is somewhat difficult, especially in the early grades. One way to interpret this quantity is to say that for these data, the closest we can come to having a fair share value is for seven of the games to have a score of 5 and the two remaining games to have a score of 4.

This process of using cubes to determine the fair share value mirrors the algorithm for finding the mean: Combining all the goals at the outset maps to adding all the data points. Distributing the goals one at a time to each of the nine games until they are all accounted for maps to dividing by the number of data points. The cubes provide visual representations of both the fair share value and the process of finding the mean. Eventually, elementary students learn that the fair share value and the mean represent the same quantity for a collection of data.

4. Interpret the Data.

At the end of the statistical problem-solving process, students reflect on their data-gathering procedure and interpret their results. The group might consider the following questions:

- "Would we expect the score from every game to be exactly the same next weekend? Why, or why not?"
- "Do you think the fair share value would still be the same if we did this same activity next Monday?"
- "What if we collected data from games involving high school teams instead of our teams?"
- "What if we collected data early in the season or late in the season? Would we expect different results?"

All these questions are intended to focus students' thinking on the issue of *difference in distributions of data* and what contributes to *variation* in the data distributions. Most of these questions do not have clear-cut answers; the objective is not to find *the* answer but for students to pose various factors that could influence the data.

Extensions

This activity can be extended by reversing the process for determining the fair share value. Suppose we know that the fair share value for nine games is 6. How many total goals might have been scored in each of the nine games if the fair share score is 6? To allow students to explore this independently, provide them with cubes and let them randomly make up the scores for nine games. You might put restrictions on some of the groups, such as "None of the games had a score of 6" or "Two of the games had a score of 10." The simplest distribution is for each of the nine teams to score 6 goals (see fig. 14.7). Figure 14.8 shows two other possibilities.

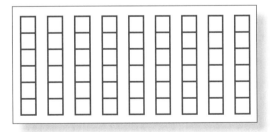

Fig. 14.7. Nine team scores, all size 6

Curriculum Issues in an Era of Common Core State Standards for Mathematics

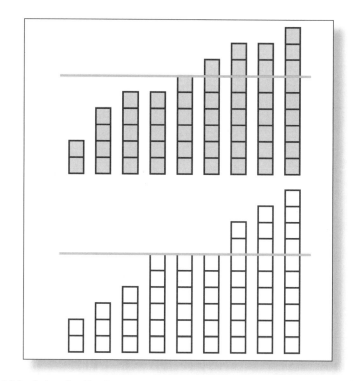

Fig. 14.8. Cube distributions with horizontal gray lines representing 6 goals

A question could arise about the two sets of scores in figure 14.8: "Which collection is 'closer' to being fair?" In statistical terms, the question is equivalent to asking, "Which of the two sets of total goals varies when compared with the mean?" Our objective is to develop a quantity that measures "how close" each of the cube distributions is to being fair. Many students will identify the white cube distribution as the one that is closer to fair because it has more sixes than the black cube distribution. Another approach can be developed by examining how many cubes must be moved to level off the two distributions at the fair share value of 6. In general, "a step" occurs when a cube is moved from a tower *above* the fair share value of 6 to a tower *below* the fair share value of 6. To determine how close a snap cube representation of data is to being fair, we can count how many steps it takes to make it fair. The fewer steps it takes to make the distribution fair, the closer the distribution is to being fair. In the example in figure 14.8, the black cube distribution is closer to being fair because it takes eight steps to level the towers, whereas the white cube distribution requires nine steps to make it fair. The next chapter in this section, by Kader and Mamer, takes this a step further by attaching values to each step.

Conclusion

Elementary school students (or those working at level A as described in the GAISE report) should be exposed to both categorical and discrete quantitative data. Students should be actively involved in the statistical problem-solving process: designing the questions to answer, collecting and representing the data, analyzing the data, and interpreting the data. Teachers should plan for questions that will help push students to interpret the collected data. "What if" questions help students begin to understand the nature of variability, a fundamental concept in data analysis. In summary, students completing level A should understand—

- the idea of the distribution for a set of data and how to represent and summarize the distribution (categorically or numerically);
- the concept of fair share as the mean value for a set of numeric data;
- the algorithm for finding the mean; and
- the notion of "number of steps" to obtain fairness as a measure of variability about the mean.

The fair share, or mean value, provides a basis to compare between two groups of numerical data with different sizes (the group total is not an appropriate comparison when sample sizes differ). Students explore this concept at level B (typically in the middle grades). Also at level B, students transition from conceptually viewing the mean as a fair share value to the mean as a *balance point* of a distribution and extending the mean as a numerical summary for continuous numerical data. The analysis of categorical data is nicely extended at level B by incorporating a student's new understanding of proportional reasoning to easily compare groups (not necessarily of the same sample sizes). Kader and Mamer explore these and other related ideas in the next chapter in this section.

References

College Board. *College Board Standards for College Success: Mathematics and Statistics*. New York: College Board, 2006.

Common Core State Standards Initiative. *Common Core State Standards for Mathematics*. Washington, D.C.: National Governors Association Center for Best Practices and the Council of Chief State School Officers, 2010. hhtp://www.corestandards.org.

Franklin, Christine, Gary Kader, Denise Mewborn, Jerry Moreno, Roxy Peck, Mike Perry, and Richard Scheaffer. *Guidelines for Assessment and Instruction in Statistics Education (GAISE) Report: A Pre-K–12 Curriculum Framework*. Alexandria, Va.: American Statistical Association, 2007.

Kader, Gary, and Jim Mamer. "Statistics in the Middle Grades: Understanding Center and Spread." In *Curriculum Issues in an Era of Common Core State Standards for Mathematics*, edited by Christian R. Hirsch, Glenda T. Lappan, and Barbara J. Reys, pp. 175–183. Reston, Va.: National Council of Teachers of Mathematics, 2012.

National Council of Teachers of Mathematics (NCTM). *Curriculum and Evaluation Standards for School Mathematics*. Reston, Va.: NCTM, 1989.

———. *Principles and Standards for School Mathematics*. Reston, Va.: NCTM, 2000.

Chapter 15

Statistics in the Middle Grades:
Understanding Center and Spread

Gary Kader
Jim Mamer

THE GAISE REPORT (Franklin et al. 2007) emphasizes the importance of students having experience with statistical thinking throughout the pre-K–12 curriculum. The continuing attention to statistics across the school mathematics curriculum is also evidenced in the Common Core State Standards for Mathematics (Common Core State Standards Initiative 2010).

According to the GAISE proposed learning progressions of key ideas, students' encounters with statistics in the middle grades should build on their foundational experiences from the elementary grades and provide a link to the inferential types of statistical thinking developed at the high school level. Middle-grades students should be actively involved in the statistical problem-solving process described in the GAISE report. That process involves (1) formulating a question that can be addressed with data, (2) collecting data to address the question, (3) analyzing the data, and (4) interpreting the results.

As students transition between levels A and B, they begin to rely less on data representations that display individual outcomes (e.g., line plots, dot plots, picture graphs) and to use representations based on groupings of the data and numerical summaries of data (e.g., histograms and box plots). From these representations, new questions arise: "Where is the center of the data located?" "How spread out are the data?" "Where do the data cluster?" Many representations for numerical data developed at level B focus on central location and the spread (amount of variability) in the data.

The Mean as the Balance Point: Variation in Data from the Mean

In the first chapter in this section, Franklin and Spangler describe the mean as the "fair share" value for a collection of discrete numeric data and the "number of steps to fair share" as a measure of variation in the data that are developed for level A. This chapter presents level B examples that build on the ideas in the first chapter, expands the notion of the mean to include the concept of a balance point for the data distribution, and develops an alternative measure of variation about the mean. (A similar activity is described in the GAISE report [Franklin et al. 2007].)

Adapted from Kader, Gary, and Jim Mamer. "Statistics in the Middle Grades: Understanding Center and Spread." *Mathematics Teaching in the Middle School* 14 (August 2008): 38–43.

The dot plots (line plots) in figures 15.1a and 15.1b show data from the Franklin chapter for two sets of scores for nine soccer games. Each set of scores has a mean value of 6. The number within each circle represents the *distance* each score is from the mean score. Notice in each plot that the total distance from the mean is the same for the values above the mean as for the values below the mean. For figure 15.1a, the total distance from the mean for the values both below and above the mean is 8. For figure 15.1b, this total distance is 9. The distances for the values below the mean are *balanced* by the distances for the values above the mean. This example shows why the mean value for a collection of numeric data indicates the balance point of the data distribution as well as the center of the distribution.

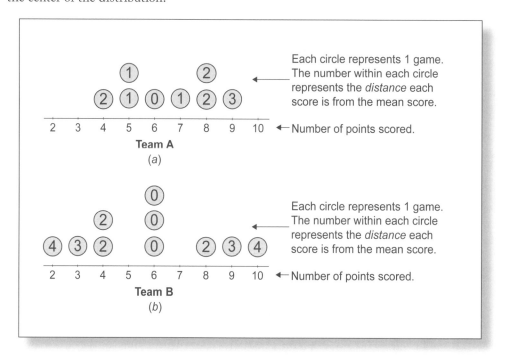

Fig. 15.1. Total points scored in each of nine games

The two distributions shown in figures 15.1a and 15.1b have the same mean. In comparing the two distributions, we can ask, "In which distribution do the data vary more from the mean?" One way to assess the amount of variability in data from the mean is to find the total distance between all data values and the mean. These distances are all considered to be positive, and their total is called the *sum of the absolute deviations* (SAD) from the mean. For the data in figure 15.1a, the SAD is 16; for the data in figure 15.1b, the SAD is 18. Thus, according to the SAD, there is more variation in the data in figure 15.1b than in figure 15.1a.

The data in figure 15.2a differ more from the mean than do the data in figure 15.2b; however, the SAD for both sets of data is 8. The SAD in figure 15.2b is based on eight data values but only two values in figure 15.2a. Dividing the SAD by the number of values gives the mean of the absolute deviations (MAD), which adjusts for the difference in group sizes: MAD = SAD/(number of values).

Statistics in the Middle Grades

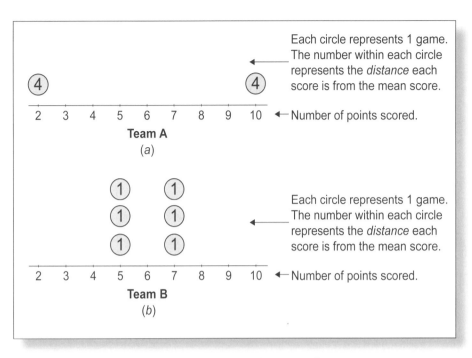

Fig. 15.2. Total points scored

The MAD for figure 15.2a is 4, whereas the MAD for figure 15.2b is 1. Thus, the data in figure 15.2a differ on average by 4 from the mean, whereas the data in figure 15.2b differ on average by 1 from the mean. The MAD provides a measure of average variability in the data compared with the mean and is a precursor for the standard deviation, the more commonly used measure for the amount of variability in data from the mean. The standard deviation builds on the idea of looking at the distance from each data value to the mean. However, instead of averaging the absolute deviations, the squared deviations are averaged. This quantity, called the *variance*, is in squared units, and the standard deviation is the square root of the variance. It is recommended that the standard deviation be developed at level C. (See Kader 1998 for more information on this idea.)

Memorizing Words: A Comparison Problem

Another useful way of summarizing center and spread in numerical data uses a five-number summary (minimum value, first quartile, median, third quartile, and maximum value). These markers form four groups with approximately 25 percent of the data in each group. The box plot, also called a box-and-whisker plot, is a graphical representation of the data distribution based on the five-number summary. Together, they provide both visual and numerical information about—

- the *center* of the distribution, based on the location of the median within the central box of the graph;
- the *spread* in the middle 50 percent of the distribution, based on the width of the box (the interquartile range); and
- the overall spread of the distribution, based on the range.

Problems involving comparisons of two or more distributions are common in statistics, and box plots are especially useful in making such comparisons. Many states' mathematics curriculum guidelines call for box plots to be introduced in the middle grades.

The statistical task that follows was adapted from a lesson described in the Learning Math Project (WGBH Educational Foundation 2001). The task engages students in exploring how easy or hard it is to memorize words in lists. Data are summarized from two seventh-grade classes in Ohio. Both classes had completed a three-week unit on statistics that introduced students to box plots. The teacher launched the activity by asking students in small groups to describe situations in which they had to memorize some kind of text and the strategies they used. Included among the examples that students gave were studying, spelling, and playing video games. The teacher displayed the following two lists and asked, "Which of the lists, A or B, do you think would be easier for most people to memorize?"

List A Words	List B Nonwords
BOSTON	MZAPDR
EAR	CTG
CART	OXCS
BUG	AEA
PAPER	SKEOC

The class conjectured that list A would be easier to memorize. They became intrigued with finding out whether they were correct and how different for other students was the experience of remembering the two lists.

Asking a Statistics Question

The class formulated the following question:

> Do people tend to score higher on list A than on list B?

Students then had to think about how to design a statistical investigation to study the question. The class discussion focused on how to *measure* someone's ability to recall words or nonwords. After an engaging discussion, the following experiment was agreed on:

- List A will be words, and list B will be nonwords. Each list will include twenty "words," each with exactly three characters.
- Each student will be assigned a list and have two minutes to study it. After a thirty-second pause, each student will have two minutes to write down as many "words" from the list as he or she can recall.

Students discussed how to score the lists of "words" that were recalled, with several suggesting scoring methods that penalized students for listing words that were not on the original list. However, the final class decision was not to invoke a penalty.

Collecting Data to Address the Question

After a statistics question has been developed, data can be collected. Several students pointed out that some people are better at memorizing than others, and that it would be unfair if all those students ended up with the same list. They decided to form two balanced groups of students; each group would have some who were more proficient and others who were less proficient with memorization. Because those students are not known in advance, the class decided to randomly assign students to the two groups, hoping that randomness would provide two balanced groups. Introducing randomness in data collection is a transition for students from level A to level B. The use of randomness in data collection links probability to statistics and is more fully developed at level C. In the next chapter, Scheaffer and Tabor discuss the use of randomness in data collection and its role in statistical inference.

In the two classes, nineteen students were assigned to list A and seventeen to list B. After separating into the two groups, the students carried out the procedure described earlier. The results were collected, scored by the teacher, and presented to the classes the next day.

Part 1: Analysis and Interpretation

In the initial analysis of the data, students constructed comparative dot plots by hand (fig. 15.3 shows a computer-generated plot). Students' comments during discussion of the dot plots included these:

Joey: Forty-seven percent of the list A kids did better than the 10.

Samantha: The data is clustered and shifted more high on graph A.

Josh: The minimum and maximum on graph A are higher than on B.

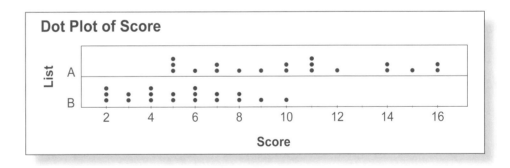

Fig. 15.3. Comparative dot plots for scores

After the discussion, the teacher posed several specific questions based on the dot plots. Students worked in groups to create arguments about whether people were more successful when working with list A than with list B. The questions and sample student responses are shown in table 15.1.

Table 15. 1
Questions Related to Dot Plots and Samples of Students' Responses

Question Number	Question	Samples of Students' Responses
1	Looking at the dot plots, did all students who studied list A score higher than all the students who studied list B? Explain.	No, because list B had numbers 5–10 and in list A, people had also scored 5–10. No, only nine people in list A scored higher than the highest score in list B.
2	Looking at the dot plots, did most students who studied list A score higher than students who studied list B? Explain.	Yes, most did better, but not all of them. About half of them scored better than them, 47 percent. No, because only 47 percent scored above the highest score on list B.
3	Looking at the dot plots, give three statements (based on the plots) that suggest that students who studied list A generally scored higher than students who studied list B.	Minimum of list B was 2. Minimum of list A was 5. List A did not go from 2 to 4 but B has 35 percent from 2–4. Forty-seven percent scored higher than list B; 53 percent of list A were from 5–10.
4	In what ways are the dot plots different? Explain how they are different.	List A is more spread out, whereas list B is more clumped together.

Students' responses to question 1 were generally very good. They recognized that although students generally did better on list A, there is some overlap in the dot plots. Responses to question 2 also were quite good. Many students identified that 9 of 19 (47 percent) scores on list A are above the highest score on list B. Questions 3 and 4 were less specific, and students found them to be more challenging. The responses varied more. In question 4, many students recognized the greater variation in list A, but no students compared the ranges.

Part 2: Analysis and Interpretation

Next, the teacher asked the students to determine the five-number summary for each list and to construct comparative box plots by hand (a computer-generated plot is shown in fig. 15.4). They returned to the statistical question they were trying to answer.

Do people tend to score higher on list A than on list B?

The groups responded to three questions that were based on the box plot analysis. The questions and the samples of students' responses are shown in table 15.2.

Statistics in the Middle Grades

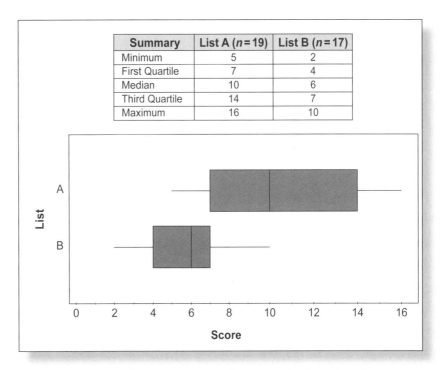

Fig. 15.4. Five-number summaries and comparative box plots for scores

Table 15.2
Questions Related to Box Plots and Samples of Students' Responses

Question Number	Question	Samples of Students' Responses
1	Looking at the box plots and five-number summaries, did all students who studied list A do better than students who studied list B? Explain.	No, because if you put both box plots on top of each other they would overlap, meaning somebody from both box plots scored the same. No, because list B goes up to the median on list A, which means half of the students had done the same as some of the students on list B.
2	Looking at the box plots and five-number summaries, give three statements (based on the plots) that suggest that the students who studied list A did better.	List A's median is higher than list B's median by 40 percent. About 50 percent on list A did better than the highest score on B.
3	In what ways are the box plots different? Explain how they are different.	List A has a larger box than list B's.

Students generally had more difficulties addressing the questions with box plots than they did with dot plots. As with the dot plots, the goal in question 1 is to recognize that some overlap exists in the box plots. For question 2, many students recognized that the median for list A (10) is higher than the median for list B (6). Many students also used statements involving proportional reasoning in addressing question 2. Several students compared quartiles and extremes and pointed out that these summaries are all higher for list A. On question 3, a few students recognized that the box plots suggest more variation in the scores from list A than from list B. Ideally, students would support this observation by comparing and contrasting summary measures of variation, in this case, the ranges and the interquartile ranges. As students gain experience with statistical problem solving, the goal is for them to become adept at connecting numerical summary measures to graphical representations of data.

Students and the teacher were extremely positive about this statistical investigation. The activity allowed students to be involved with all four components of the statistical problem-solving process. During the activity, as students discussed what they saw in the plots, they used statistical terms such as *clusters, spread, minimum, maximum, median, most data,* and *percentages.* By working together, they were able to feed off one another's statistical ideas. This activity illustrates how authentic problems can be used to develop statistical concepts. Such activities take students beyond simply calculating summary measures and constructing plots and push them to think about how to use the summary measures and plots to address various questions.

Although the box plot is a relatively simple graph to construct, Bakker, Biehler, and Konold (2004) point out that middle-grades students generally have difficulties interpreting box plots. One explanation is that the four groups of data formed from the five-number summary are based on percentages and that their interpretation requires proportional reasoning. Guidance and practice are required for middle-grades students to understand the five-number summary and be able to interpret and compare box plots. Middle school is an appropriate place for this effort to begin, because developing proportional reasoning skills is a goal for the middle and junior high grades.

Conclusion

The GAISE report describes three developmental levels of evolving statistical concepts. Experiences at level B should provide the link between the more concrete experiences from level A to the inferential types of thinking developed at level C. Two contexts for developing student thinking about the center and spread for discrete numerical data are explored. The representations discussed in these activities also are appropriate for continuous numerical data. These statistical tasks promote student understanding of various numerical summaries and illustrate connections between various representations. Both tasks build on ideas developed at level A, promote proportional reasoning at level B, and provide the foundation for statistical thinking developed at level C.

References

Bakker, Arthur, Rolf Biehler, and Cliff Konold. "Should Young Students Learn about Box Plots?" In *Curricular Development in Statistics Education,* edited by Gail Burrill and Mike Camden, pp. 163–73. Voorburg, Netherlands: International Statistics Institute, 2004.

Common Core State Standards Initiative. *Common Core State Standards for Mathematics.* Washington, D.C.: National Governors Association Center for Best Practices and Counsel of Chief State School Officers, 2010. http://www.corestandards.org.

Franklin, Christine, Gary Kader, Denise Mewborn Jerry Moreno, Roxy Peck, Mike Perry, and Richard Scheaffer. *Guidelines for Assessment and Instruction in Statistics Education (GAISE) Report: A Pre-K–12 Curriculum Framework.* Alexandria, Va.: American Statistical Association, 2007.

Franklin, Christine A., and Denise A. Spangler. "Statistics in the Elementary Grades: Exploring Distributions of Data." In *Curriculum Issues in an Era of Common Core State Standards for Mathematics,* edited by Christian R. Hirsch, Glenda T. Lappan, and Barbara J. Reys, pp. 165–173. Reston, Va.: National Council of Teachers of Mathematics, 2012.

Kader, Gary. "Means and MADS." *Mathematics Teaching in the Middle School* 4 (February 1998): 398–403.

Scheaffer, Richard, and Josh Tabor. "Statistics in the High School Mathematics Curriculum: Building Sound Reasoning under Uncertain Conditions." In *Curriculum Issues in an Era of Common Core State Standards for Mathematics,* edited by Christian R. Hirsch, Glenda T. Lappan, and Barbara J. Reys, pp. 185–193. Reston, Va.: National Council of Teachers of Mathematics, 2012.

WGBH Educational Foundation. *Learning Math: Data Analysis, Statistics, and Probability.* Boston, Mass.: WGBH, 2001.

Chapter 16

Statistics in the High School Mathematics Curriculum:
Building Sound Reasoning under Uncertain Conditions

Richard Scheaffer
Josh Tabor

STATISTICS is the key to decision making in the information age. The importance of statistical thinking for life and work is reflected in state and provincial mathematics frameworks and in national standards. These documents all place statistics (along with number, algebra, and geometry) as one of the "big four" strands in the mathematical sciences. But statistics, often described as data analysis, is the strand least emphasized, for many reasons—lack of teacher preparation on content and pedagogy, lack of good instructional materials, lack of emphasis on standardized tests, and lack of models for integrating statistical thinking into the school mathematics curriculum. To provide guidance on what and how to teach statistics effectively as part of the K–12 mathematics curriculum, the American Statistical Association (ASA) developed *Guidelines for Assessment and Instruction in Statistics Education* (GAISE) (Franklin et al. 2007). This chapter, along with chapters 14 and 15 by Franklin and Spangler and by Kader and Mamer, respectively, illustrate how GAISE can be used to create a K–12 trajectory of the development of basic ideas related to the distribution of a variable based on data. These ideas are built by viewing statistics as an investigative process and by incorporating the guiding principles for teaching outlined in figure 16.1.

Distributions summarizing quantitative information are characterized through the concepts of *shape, center,* and *spread*. The topics covered in this chapter assume that students will have been exposed to the calculation and interpretation of basic measures of center (e.g., median as the middle value and mean as a fair share or balance point) and spread (e.g., interquartile range and standard deviation) and how uses of these measures relate to the shape of a distribution and the purpose of a study. The specific goal of this chapter (paralleling level C of the GAISE report) is to discuss the teaching of distributions of commonly used sample statistics, such as means and proportions, and how these

---

Adapted from Schaffer, Richard, and Josh Tabor. "Statistics in the High School Mathematics Curriculum: Building Sound Reasoning under Uncerain Conditions." *Mathematics Teacher* 102 (August 2008): 56–61.

> **Steps in the statistical problem-solving process**
> - Formulate questions
> - Collect data
> - Analyze data
> - Interpret results
>
> **Guiding principles for teaching statistics**
> - Conceptual understanding takes precedence over procedural skill.
> - Active learning is key to the development of conceptual understanding.
> - Real-world data must be used wherever possible in statistics education.
> - Appropriate technology is essential in order to emphasize concepts over calculations.
> - All four steps of the investigative problem-solving process should be encountered at each grade level.
> - The illustrative investigations should show situations in which the statistics is essential to answering a question, not just an add-on.
> - Such investigations should be tied to the mathematics they illustrate, motivate, and emphasize.
>
> Source: GAISE Report (Franklin et al. 2007)

Fig. 16.1. Steps in the statistical problem-solving process and guiding principles for teaching statistics

distributions are used in inferential (i.e., plausible) reasoning. These statistics can be derived from data collected through a survey with random sampling or a randomized experiment.

A Sample Survey

One of the most straightforward, popular, and potentially interesting ways to introduce basic concepts of inferential reasoning is through a sample survey. Students should be cautioned, however, that a good sample survey is based on a random sample from a well-defined population, which is easy to conceptualize but difficult to achieve in practice. The following example brings to light the key ideas of the statistical problem-solving process in the context of collecting and analyzing data from a sample survey.

Asking a Statistics Question

Students in a high school mathematics class decided to study the strictness of parents or guardians of students in the school. They realized that, because the meaning of *strict* is ambiguous, they could not simply ask as a survey question, "Are your parents or guardians strict?" After some debate, they decided on a variety of survey questions that might get at the idea of strictness. One of these was "Do you have a curfew?" From this they formulated the following statistics question: "What percentage of students in the school have a curfew?"

Collecting and Analyzing Appropriate Data

A random sample of 100 students was selected from the approximately 1,200 students in the school, and all sampled students responded to the survey. The results (table 16.1) showed that 53 of the 100 reported having a curfew. Data collected through this survey question are categorical; "yes" meant that the student did have a curfew, "no" meant that the student did not. However, the sample summary statistic (proportion in sample with a curfew = 0.53) is numerical.

Table 16.1
Summary of Data on the Curfew Question

	Had a Curfew	Did Not Have a Curfew	Total
Females	33	35	68
Males	20	12	32
Total	53	47	100

Interpreting Results

Students should understand that the true proportion of all students in the school is not likely to be 0.53 and that another sample of 100 students might very well produce a different result. This is the underlying idea of a sampling distribution: the sample proportion will vary from one sample to another. How much variation can we expect? Time for technology! If we assume that the true population proportion is 0.53, students can simulate the repeated selection of samples of size 100 from a "theoretical" population that has a 53 percent chance of generating a favorable outcome. An easy way to do this is to sample 100 two-digit random numbers, with the numbers 01 through 53 representing yes and the numbers 54 through 99 as well as 00 representing no. This sampling can be done by using a table of random digits, a computer (using software or websites such as www.random.org), or a graphing calculator (on a TI-83/84, use Math: Prb: 5RandInt(00,99)). Figure 16.2 shows a simulated sampling distribution of sample proportions from 200 such samples. This sampling distribution has a mean of 0.53 and a standard deviation of 0.05 and is nicely represented by the normal distribution (overlaid smooth curve) with that same mean and standard deviation.

For a normal distribution, about 95 percent of measurements lie within two standard deviations of the mean. This measure of two standard deviations on either side of the mean, sometimes called the margin of error, forms an interval of reasonably likely outcomes. The margin of error is 2(0.05) = 0.10 for this example and would not change much even if the assumption of 0.53 for the true population proportion is off a bit. Finally, the plausible values of the population proportion that could produce such a sample outcome as reasonably likely must lie between 0.43 and 0.63 (0.53 ± 0.10), as displayed

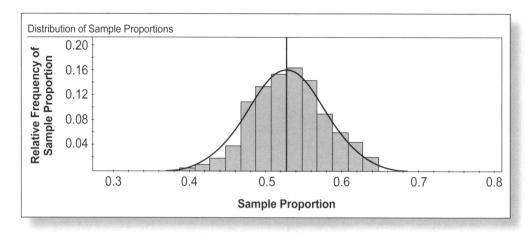

Fig. 16.2. Simulated sampling distribution for a sample proportion

in figure 16.3. The two normal distributions sketched here are centered at 0.43 and 0.63 and have the same standard deviation (0.05). Each has the observed proportion of 0.53 at an edge of its reasonably likely outcomes. Thus, any normal distribution with nearly the same standard deviation that is located between these two distributions would also have the observed proportion of 0.53 within its reasonably likely outcomes. The interval of plausible values, 0.43 to 0.63, is called a 95 percent confidence interval.

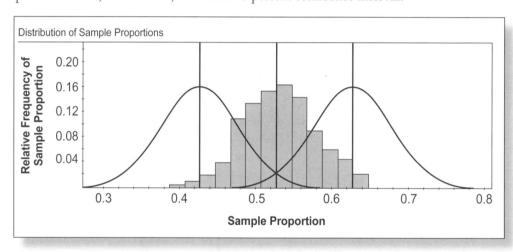

Fig. 16.3. Interval of plausible values for a population proportion

One statistical analysis almost always leads to another. Because the observed proportion of the surveyed students having curfews was higher for males than for females, many students jumped to the conclusion that this proportion must be true for the school population as a whole. Table 16.1 shows that the sample proportions of those having curfews were 20/32 = 0.625 for males and 33/68 = 0.485 for females. Remembering that sample proportions have inherent variation and that the sample size here is only 100, some students began to question whether this difference would hold up for the population of the whole school.

How can a statistical question that gets at this issue be formulated? The key lies in the randomness built into the sample selection process. Given that there are 68 females and 32 males in the sample and if we assume that a randomly selected male and a randomly selected female are equally likely to have a curfew, can the division seen in the observed sample (in which 33 females and 20 males have a curfew) be attributed simply to chance? This is a probability question that can be answered through simulation.

The simulation can be done manually or by using technology, but the principle is as follows. Gather 100 chips and mark 53 of them "curfew." Randomly split the 100 chips into piles of 68 "females" and 32 "males"; the randomization ensures that each "curfew" chip has an equal chance of ending up in either pile. Record the number of "curfews" among the "males." Repeat the process many times and plot the resulting distribution, now called a randomization distribution because it depends only on rerandomizing the observed data. Figure 16.4 shows the result for one such simulation of 200 trials. Here, 20 or more males with curfews out of a sample of 32 males occurs 25 times out of 200 trials, or 12.5 percent of the time. In most statistical investigations, this would be considered fairly often, so there is little evidence to support a contention that the males in the school have a higher proportion of curfews. The split seen in the sample could have been due simply to the chance mechanism of random sampling. (*Teaching note*: Instead of chips, you can also use regular playing cards, colored beads, or any other objects of the same size that can be identified as curfew or noncurfew and randomly divided into two groups [male and female].)

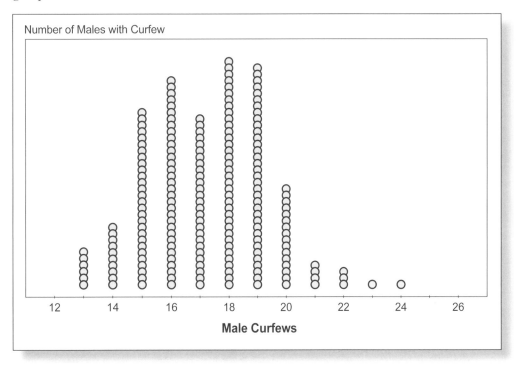

Fig. 16.4. Randomization distribution for the number of males with curfews

Another question: Is the 68-to-32 split between females and males a bit unusual in a random sample of 100? This, too, can be verified by randomization.

A Comparative Experiment

The GAISE report and the Common Core State Standards for Mathematics (Common Core State Standards Initiative 2010) specify that high school students should understand the basic ideas behind both sample surveys and experiments. The following investigation illustrates some of the similarities and differences between surveys and experiments and incorporates an element of geometry as well.

The task is to approximate the area of a triangle. Each student in a class is asked to approximate the area through one of two methods: by using the area formula (area = [1/2] · base · height [method 1]) or by estimating through the use of graph paper (method 2).

One foundation of a good experiment is the random assignment of the methods (treatments) to students (experimental units). Assigning all the careful students to one method and all the careless ones to the other would *not* be the basis for a good experiment. Random assignment helps balance these characteristics (and any others that might affect the results) in the two treatment groups.

After assigning the treatments and collecting the data, have the students plot the data to see what they reveal. The original data are plotted in figure 16.5, from which it is quite obvious that four areas calculated by means of the formula are far from all the other calculations. Further checking revealed that these four students calculated in centimeters rather than inches; the corrected data are shown in figure 16.6. Being alert to outliers in a data set is strongly recommended, because these are often errors of measurement but on occasion may represent unique discoveries.

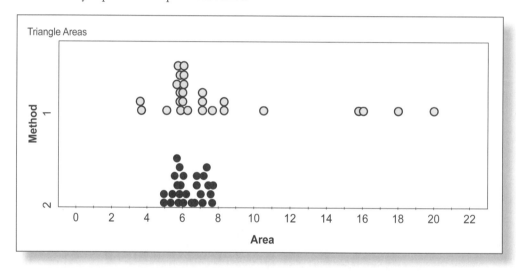

Fig. 16.5. Original area data (method 1 shown at top, method 2 shown at bottom)

Class discussion focusing on shape, center, and spread (see table 16.2 for summary statistics) should conclude that the two methods for computing the area of a triangle are about the same with respect to center but that the formula method may reflect greater variability. Even though one distribution is slightly skewed, using the sample mean as the measure of center is reasonable. The question on centers can then be phrased in terms of the difference between means: Can the difference between the sample means (observed to be about 0.05 square inch) be attributed to the randomization alone, or should we look for another cause?

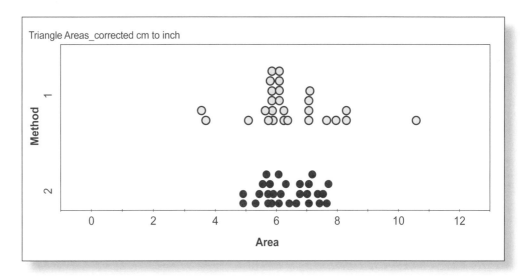

Fig. 16.6. Corrected area data (method 1 shown at top, method 2 shown at bottom)

Table 16.2
Summary Statistics for the Corrected Area Data

	Method 1	Method 2
Number	27.00	8.00
Mean	6.35	6.30
Median	6.00	6.16
Standard deviation	1.40	0.83

If both methods worked equally well, we would expect no difference in the sample means. However, because randomization cannot guarantee that the two groups are equivalent, the sample means will differ somewhat. To investigate whether the observed difference could be explained by the randomization process alone, we repeat that process many times so as to generate a simulated randomization distribution of the difference in sample means. If the observed difference is too big to be accounted for by chance alone, then a real difference in the methods is a more plausible explanation than is attributing the observed difference to the randomization.

Figure 16.7 provides a simulated randomization distribution of 200 trials each for the differences in means for the triangle area data. This distribution was produced by repeatedly splitting the 55 areas at random into groups of 27 for method 1 and 28 for method 2 and calculating the statistic of interest each time. The observed difference in means of 0.05 is almost exactly at the center of the rather symmetric distribution of the difference in means. So this value could well be due to the students' randomization pattern rather than to any real difference in methods. (*Teaching note:* To perform this simulation by hand, students can write each of the 55 areas onto equal-sized slips of paper, mix them up, and divide them into two piles: 27 for method 1 and 28 for method

2. Alternatively, students could use the Randomization Distribution feature of CPMP-Tools software (www.wmich.edu/cpmp/CPMP-Tools/), described in Hart, Hirsch, and Keller (2007), or a TI-83/84 graphing calculator.

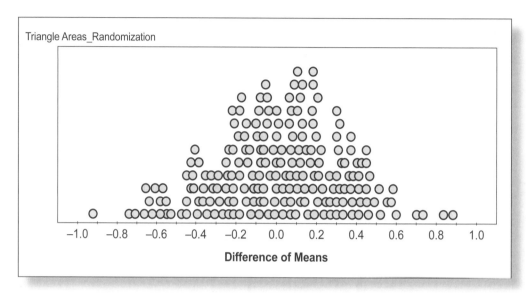

Fig. 16.7. Randomization distribution of difference in means

The fact that data from both methods center around 6.3 square inches is strong evidence that this number is a good approximation of the true area of the triangle. If a difference in means were evident, further investigation of the accuracy of the methods would be in order. The only difference in the methods, then, seems to be in the variation of the data, with method 1 being the more variable. With an appropriate measure of variation, such as the standard deviation or the mean absolute deviation as developed in the Kader and Mamer chapter, comparisons of spreads can be made by randomization as well.

Conclusion

Many interesting questions that can be posed in the classroom can be addressed through the methods presented in this chapter. For example, in a sample survey, students could investigate the question "Do boys or girls have more pairs of shoes?" In a comparative experiment, students could investigate the question "Does the amount of homework assigned have a significant effect on test results?" Even teachers can do their own research: "Do students who sit in the front of the classroom perform better than students who sit in the back?" In this last example (as in all experiments), random assignment of students is essential, because more motivated students may choose to sit in front.

Most often, statistics in high school mathematics is either embedded as part of algebra or set up as separate units in an integrated mathematics program. In either setting, statistics can vitalize the mathematics being taught because students tend to be inherently interested in data on practical problems relating to their lives and the world around them. Data distributions are built on real number lines and make use of proportional reasoning (e.g., relative frequency). Measures of center and spread can be formulated algebraically, and algebra can be used to show the effect of linear transformations on these measures (in changing from centimeters to inches, for example). The patterns that emerge in sampling and randomization distributions are the result of the laws of probability. However, the essentials of statistical reasoning must be the main focus in bringing statistics to life. And mathematics is a necessary underpinning to that reasoning.

References

Common Core State Standards Initiative (CCSSI). *Common Core State Standards for Mathematics*. Washington, D.C.: National Governors Association Center for Best Practices and the Council of Chief State School Officers, 2010. www.corestandards.org.

Franklin, Christine, Gary Kader, Denise Mewborn, Jerry Moreno, Roxy Peck, Mike Perry, and Richard Scheaffer. *Guidelines for Assessment and Instruction in Statistics Education (GAISE) Report: A Pre-K–12 Curriculum Framework*. Alexandria, Va.: American Statistical Association, 2007.

Hart, Eric W., Christian R. Hirsch, and Sabrina A. Keller. "Amplifying Student Learning in Mathematics Using Curriculum-Embedded Java-Based Software." In *The Learning of Mathematics*, Sixty-ninth Yearbook of the National Council of Teachers of Mathematics (NCTM), edited by W. Gary Martin and Marilyn E. Strutchens, pp. 175–204. Reston, Va.: NCTM, 2007.

Section VII
Improving Vertical Articulation: Challenges and Promising Practices

Curriculum Issues in an Era of Common Core State Standards for Mathematics

Introduction

The beginning of every school year brings challenges to students from a variety of sources. This year's teachers may structure their classrooms differently from last year's. Prior to transitioning to the Common Core State Standards for Mathematics (CCSSM), which includes grade-level standards for students in grades K–8, teachers at the same grade may have emphasized different topics. Teachers' expectations for how students participate in instructional activities may differ. Students also face the problem of forgetting what they learned last year, especially if they have not used particular mathematical knowledge or skills over the summer months. Such annual challenges are minor compared with those that students face during crucial transition points—from elementary school to middle school, middle school to high school, and high school to postsecondary education. The three related chapters in Section VI take on these challenges.

A theme running across the three chapters in this section is the importance of teachers frequently communicating within and across grade levels in making decisions about when and how to develop important mathematical ideas. Teachers' communication is especially crucial in this period of interpreting and implementing the CCSSM, in which many topics appear at lower grades than has been common practice. Such collaboration can also help build a common classroom culture from year to year. When teachers agree on where they will focus instructional time and on what they expect of students in the classroom, then students can spend more time each year focusing on important mathematical topics.

The first chapter identifies the challenges facing students transitioning from elementary to middle school. It offers practical suggestions to overcome those challenges, including teachers visiting other classrooms and having comparative discussions of assignments and students' work. The second chapter offers specific suggestions on easing students' transition from middle to high school, with particular attention not only to content and instructional alignment but also to psychological and social factors to consider. The final chapter focuses on the crucial transition from high school to postsecondary education. The authors focus not only on the challenges but also on specific actions teachers can take to ease students' transitions and specific messages that teachers should share with their students.

As you read these three chapters, think about what structures your school district could put in place to improve articulation and students' transitions from one grade or level to the next as you begin implementing the CCSSM.

Questions for Reflection and Collective Discussion

1. What practices have teachers in your district found successful in improving content and instructional articulation from one grade to the next? From one grade band to the next? What challenges remain, and how might you and your colleagues address them?

2. What difficulties, if any, have you found in trying to collaborate with other teachers to improve coordination and articulation within or across grades? How might you address those difficulties?

3. What successes have you had in collaborating with others to improve articulation across grades, in both content and instructional practices, including assessment? How were those efforts initiated or supported?

4. How are the challenges of transition from elementary to middle school similar to, and different from, those from middle school to high school?
5. How are the challenges of transition from middle school to high school similar to, and different from, those from high school to college?
6. What articulation issues will your school or district personnel need to consider as they plan for implementing the Common Core State Standards? What steps will you take to address these issues?
7. How do the curriculum materials your district uses for grades K–12 support or interfere with articulation and smooth transition from elementary to middle school? From middle to high school?
8. The roles of the teacher and students often vary from classroom to classroom, from one grade to the next, particularly at the crucial transition years. How much of an issue is this in your school building or district, and how can you address it?
9. What opportunities does your district offer teachers so that they can work together across classrooms, grades, and schools toward a coherent, well-articulated, CCSSM-oriented grades K–12 mathematics program?
10. Data show that students who did not enroll in a mathematics course during their senior year are much more likely to end up in a remedial, not-credit-bearing course on admission to college. What provisions does your district have to ensure that all college-bound students enroll in an appropriate mathematics course their senior year?

Chapter 17

Transitions from Elementary to Middle School Mathematics

Janie Schielack
Cathy L. Seeley

THE MOST critical steps to be taken now to ensure proper widespread implementation of the Common Core State Standards for Mathematics (CCSSM; Common Core State Standards Initiative 2010) are to consider phasing-in models and to launch related professional development. In Section I of this book, Confrey and Krupa recommend that phasing-in model discussions include strategies for the transition grade levels where changes in schools and schooling practices necessitate careful attention to student success (p. 12).

Transitioning from elementary school to middle school has always been a difficult time for students. A new middle school teacher describes one difference: "In elementary school, eating lunch with the teacher is a reward; in middle school, it is a punishment."

In addition to adjusting to the changes in social environment, students moving from elementary to middle school must deal with different academic expectations, reorganization of the school day, and multiple teachers (generally one per course). These challenges are exacerbated by the increased amount of new content that the CCSSM has placed in grade 6.

Previous research supplies evidence that, in general, students suffer significant declines in academic achievement in the transition from elementary school to middle or junior high school (Alspaugh 1998). In particular, students' attitudes toward—and achievement in—mathematics appear to be negatively affected in this transition (Eccles et al. 1993). Although this research does not clearly pinpoint causes of these negative effects, examining instructional materials, classroom environments, and curricula of elementary and middle schools reveals some challenges that students encounter. And, as with all challenges, these present opportunities for teachers to help students bridge this transition.

Adapted from Schielack, Janie, and Cathy L. Seeley, "Transitions from Elementary to Middle School Mathematics." *Teaching Children Mathematics* 16 (February 2010): 358–62.

Challenges

In the move from elementary to middle school mathematics, students encounter major changes in instructional materials and approaches, work expectations, school structure, and general level of difficulty in material. Teachers' awareness of these changes can help students prepare for the challenges.

The textbook and other major program materials used in a mathematics classroom establish an instructional framework. Most publishers of mathematics materials separate their products into grade-level bands of K–5 and 6–8. Some publishers provide two versions of grade 6 materials, differentiating them by whether grade 6 is situated in an elementary school or a middle school. If one thumbs through a grade 5 mathematics textbook and a grade 6 middle school mathematics textbook, even those from different programs, one gets an immediate sense of the difference in how the materials look. Color schemes differ; the grade 6 book has less white space and greater word density, a smaller font size, more word problems, and more computational items in the exercise sets.

Closer examination reveals other differences—for example, in the types of representations used. Elementary school materials usually exhibit many more types of representations than middle school materials do. Elementary school representations often are large and spacious—such as the use of regions, particularly sections of circles, to represent fractions—whereas representations in middle school materials are often more symbolic and space efficient, such as the use of linear models or the number line for representing fractions. Depending on the elementary and middle school textbooks used, these differences in representation could just as easily be reversed. The representations in the elementary textbooks might be more symbolic, and the representations in the middle school textbooks, more spacious. If teachers do not pay specific attention to the differences in instructional materials—for example, by purposefully forming connections between different types of representations—students may move from elementary to middle school with critical gaps in their understanding.

Classroom Capsule

The transition from elementary school to middle school typically results in lower achievement in mathematics.

- The challenge—How can adults help students better navigate a transition that includes new expectations, materials, instructional structures, and strategies?
- The recommendations—To improve the transition process, school personnel can:
  - Increase communication and interaction between teachers at each level.
  - Establish effective articulation and vertical alignment in the curriculum.
  - Create a supportive network for students that includes teachers, counselors, administrators, and parents.

The instructional materials chosen for different grade levels often reflect the prevalent instructional philosophies at those grades, philosophies that may also differ widely. Some programs advocate instructional philosophies that support direct instruction, the teacher as organizer and deliverer, and the student as receiver and responder. Other programs encourage students' open exploration and the teacher as facilitator. Again, these different philosophies can appear on either side of the elementary–middle school transition, but if the two sides are different, students must adjust.

How can teachers help students shift from a mathematics classroom that focused on a single, efficient problem-solving method to a classroom where problem solving builds around whole-group or small-group discussions? How will students know what they are supposed to learn in the new instructional environment? What steps can teachers take to help students learn how to engage in the discussions?

If the change is reversed, from a student-centered environment to one that is more teacher directed, what can teachers do to help students internalize questioning, observation, and inference skills and apply them in the new environment? How can teachers prepare students for new classroom rules they may encounter—for example, how and why the use of technology is or is not allowed—in either transition?

Changes in work expectations in mathematics present a huge challenge to most students transitioning from elementary to middle school. The amount of independent work expected per week, including homework, is generally greater in middle than elementary school, and a higher level of focused concentration is expected in middle school math classes. In a self-contained grade 5 classroom, mathematics instruction and skill practice can be spread throughout the day, appearing in morning warm-up exercises, at centers that are open all day, and in "sponge" activities to fill random bits of open time. Even in departmentalized grade 5 situations, math instruction is more likely to be connected to the rest of what is happening in grade 5, with classrooms close together in the same hallway.

In middle school, attention to mathematics is expected to begin at the start of math period and last until the end of that period—and this attention to mathematics is needed *only* during this period, until it is time to do homework outside school. How can teachers help students learn to compartmentalize their attention for the most efficient learning and prepare for the independent study skills expected in middle school?

Finally, entwined throughout all these environmental challenges is the increased difficulty in the mathematics content and related terminology that students encounter in middle school. In general, students progress from working mainly with whole numbers through grade 5 to spending a large amount of time in middle school working with fractions and fraction-related concepts (e.g., ratio and proportion). Middle school math teachers may use language different from that used by elementary school teachers to describe the same or closely related ideas. Middle school teachers may use a familiar word to describe a different idea (e.g., *slope* or *similar*). How can we identify where gaps exist in content development between grades 5 and 6 and develop ways to smooth the curricular transition?

Curriculum across the Grades

In examining the mathematics curriculum across grades 3–8, teachers might consider the following questions (Schielack and Seeley 2007, p. 79):

1. Which key ideas are reflected in our existing mathematics curriculum? Do NCTM's Focal Points contain any essential ideas that do not appear somewhere in our curriculum, or vice versa? If so, how do we address that deficiency?

2. Does our key-concepts sequence make sense mathematically? Does it build developmentally from grade to grade without unnecessary repetition? If not, what must be changed to make the concepts flow from grade to grade?

3. Can we determine from our curriculum where to place the emphasis at each grade? Can we tell how a topic (such as fraction operations) should be treated differently at grades 5, 6, 7, and 8 to help students learn for long-term retention without repetition each year?

4. When do we expect the most focused instruction on a particular idea to occur? Is there an appropriate amount of time for instruction in relation to the whole set of fundamental ideas at that time? If not, how can the ideas be shifted in a coordinated way to provide for that time?

5. Can we follow the development of important concepts as they move through the grades from background to foreground to background? Does that movement make sense?

Suggested Practices

As outlined in a position paper sponsored jointly by the National Middle School Association and the National Association of Elementary School Principals (March 2002), successful transition programs include the following attributes:

- **Sensitivity** to the anxieties accompanying a move to a new school setting
- **Importance** of parents and teachers as partners in this effort
- **Recognition** that becoming comfortable in a new school setting is an ongoing process, not a single event

The key to identifying and addressing the challenges that students face in transitioning from elementary to middle school mathematics is to offer opportunities for communication between grade 5 and grade 6 instructional leaders, including teachers. Communication channels can be built around direct involvement in classrooms, indirect involvement in classrooms through videotaping and sharing students' work, review and analysis of the mathematics curriculum, and counselor and parent engagement.

As the previous section outlines, elementary and middle school mathematics classroom environments differ in many ways. One of the most effective approaches to identifying essential environmental differences is through classroom visits. Teachers of grades 4 and 5 (or 6 and 7) who have the opportunity to personally observe the management and structure of elementary and middle grades classrooms can gain important insight into the challenges that students will encounter in the transition between the two environments. For example, a grade 5 teacher who visits a grade 6 math classroom might see that students are expected to enter the room, find their seats, and—as the teacher takes roll call—begin independently working on an assignment written on the board, all before the teacher addresses the class. As a result, that particular elementary teacher might plan to take a day each week for students to practice for middle school (e.g., the teacher writes a short assignment on the board, and students do the assignment without any oral directions from the teacher).

Similarly, a middle school teacher who visits an elementary school classroom could gain better understanding of some different ways students engage in challenging mathematics within a self-contained classroom environment—particularly the opportunities to connect math to their science or social studies activities.

A pair of elementary and middle school teachers might plan to visit each other's classrooms when small-group work is occurring, so that they can compare how students at the two levels are expected to work in groups. After the observations, they could discuss ways to help students transition between the different expectations.

Of course, making actual classroom visits raises many issues of proximity, schedule conflicts, time commitments, and collegial trust. For an alternative, elementary and middle school teachers can share and discuss videotapes of relevant parts of their classroom instruction. If collecting videotapes of their own instruction is impossible, elementary and middle school teachers could observe third-party videotapes of math instruction (e.g., those from the Annenberg Foundation or the Trends in International Mathematics and Science Study) and discuss how these examples are similar to or different from their own math classrooms. The most important benefit of this kind of activity is the opportunity for teachers at different grade levels to interact around teaching and learning. Third-party examples supply a common reference point for productive discussions about students' challenges as they transition from elementary to middle school math classrooms.

Mathematics teachers at different grade levels might also engage in comparative discussions of examples of assignments and students' work. If teachers observe that most assignments at the middle school level involve a much higher level of independent reading than at the elementary school level, they might identify specific ways to help students make that transition. Perhaps the elementary school teachers could infuse their math lessons with more opportunities for students to practice reading. They might share with middle school teachers some content reading techniques with which younger students are familiar—for example, underlining important words—so that the middle school teachers could plan to use these techniques early in the year to support students in further developing their independent reading skills in math. ·

Teachers could also choose specific examples of students' work to highlight writing quality and the types of representations expected at each level. For example, if middle school students are expected to show more than one way of representing a solution to a problem, then elementary school teachers might want to ensure that their students have opportunities to explore and discuss different ways of showing and explaining their answers.

A crucial component of addressing the transition from elementary to middle school math is the vertical alignment of the mathematics curriculum. Teachers in grades 3–8 provide students with the main building blocks for success in algebra, including topics that address the understanding of fractions, fraction operations, ratio, and proportion, along with important ideas related to geometry and measurement. However, for these building blocks to come together and create a strong foundation, elementary and middle school teachers must work together to help students build connections among these topics. The district's mathematics curricula must support instructional design and implementation that builds these connections. Elementary and middle school teachers could work together to build curricular bridges based on analyses of learning trajectories of key CCSSM topics in grades K–8 (Confrey, Maloney, and Nguyen 2010a, 2010b) and on examples in such materials as NCTM's *Curriculum Focal Points* (2006). Aligning major ideas across grades 3–8 efficiently addresses the curricular aspects of students transitioning from elementary to middle school mathematics.

Of course, counselors and parents make important contributions to effectively transitioning students from elementary to middle school. "School counselors, as leaders and advocates for transition programming, must inform all parents of the ways they can support their children and of the ways that they can become actively engaged in ensuring equitable placement decisions" (Akos, Shoffner, and Ellis 2007, p. 241). Parents and counselors must continue to give students a vision of what being "good in mathematics" means for a student's future. Too often, students' interests and concerns in other aspects of school and life overshadow their interest in academics.

Students will find it easier to transition between elementary and middle school math if they have supportive teachers, counselors, and parents who communicate across grade levels to align curricula and bridge differences between instructional materials and environments. Even more important, as students successfully make this transition, they also strengthen their foundation for success throughout secondary school mathematics.

References

Akos, Patrick, Marie Shoffner, and Mark Ellis. "Mathematics Placement and the Transition to Middle School." *Professional School Counseling* 10 (February 2007): 238–44. http://findarticles.com/p/articles/mi_m0KOC/is_3_10/ai_n19311518.

Alspaugh, John W. "Achievement Loss Associated with the Transition to Middle School and High School." *Journal of Educational Research* 92 (September–October 1998): 20–25.

Common Core State Standards Initiative. *Common Core State Standards for Mathematics*. Washington, D.C.: National Governors Association Center for Best Practices and the Council of Chief State School Officers, 2010. http://www.corestandards.org.

Confrey, Jere, Alan P. Maloney, and Kenny H. Nguyen. *Learning Trajectory Display of the Common Core State Standards for Mathematics, Grades K–5*. New York: Wireless Generation, 2010a.

———. *Learning Trajectory Display of the Common Core State Standards for Mathematics, Grades 6–8*. New York: Wireless Generation, 2010b.

Eccles, Jacquelynne S., Allan Wigfield, Carol Midgley, David Reuman, Douglas Mac Iver, and Harriet Feldlaufer. "Negative Effects of Traditional Middle Schools on Students' Motivation." *Elementary School Journal* 93 (May 1993): 553–74.

National Council of Teachers of Mathematics (NCTM). *Curriculum Focal Points for Prekindergarten through Grade 8 Mathematics: A Quest for Coherence.* Reston, Va.: NCTM, 2006.

National Middle School Association and the National Association of Elementary School Principals. "Supporting Students in Their Transition to Middle School." Joint position paper, March 2002. http://www.nmsa.org/AboutNMSA/Position Statements /TransitioningStudents/tabid/283/Default.aspx.

Schielack, Jane F., and Cathy L. Seeley. "Implementation of the NCTM's *Curriculum Focal Points*: Concept versus Content." *Mathematics Teaching in the Middle School* 13 (September 2007): 78–80.

Chapter 18

Transitions from Middle School to High School: Crossing the Bridge

Lisa C. Brown
Cathy L. Seeley

To ensure successful implementation of the Common Core State Standards for Mathematics (CCSSM; Common Core State Standards Initiative 2010), among the most important critical steps is to consider phasing-in models and related professional development. In Section I of this book, Confrey and Krupa recommend that phasing-in model discussions among teachers include strategies for the transition grade levels where changes in schools and schooling practices necessitate careful attention to student success (p. 12). This is particularly crucial in the transition to high school from middle school.

As students transition from middle school to high school, the academic landscape changes. This transition is complicated by several challenges—some general and some specific to mathematics—faced by students, educators, and families.

Transitioning to High School

Some of the challenges that students face when they move from middle school mathematics to high school mathematics include—

1. an insufficient alignment of mathematics instruction and curriculum across grades;
2. issues with the initial mathematics content typically taught in high school; and
3. the psychological and social factors influencing students' beliefs and perceptions about their ability to learn difficult material.

Despite these challenges, we can draw on a number of practices to help guide students through this critical transition toward greater success and opportunity in high school and beyond.

Adapted from Brown, Lisa C., and Cathy L. Seeley, "Transitions from Middle School to High School: Crossing the Bridge," *Mathematics Teaching in the Middle School* 15 (February 2010): 354–58.

Challenge: Alignment of Instruction and Curriculum

Many students who are transitioning to high school are unprepared for high school math. Among them are students who may have been comfortable in elementary mathematics yet who did not fare well through the elementary school–middle school transition and now are on a downward trajectory as they work their way through middle school (Eccles, Lord, and Midgley 1991). Lower grades that many students receive may be related to a lack of alignment of teaching strategies and expectations between elementary and middle school mathematics. Middle school teachers often face more than 150 students every day. Consequently, their mathematics students may encounter fewer opportunities for hands-on exploration and less individualized attention than they received in elementary school.

These different instructional approaches can be related to the curriculum resources used in the different grade bands. Students are likely to experience one textbook series in elementary school, an entirely different series through middle school, and yet another series in high school. This lack of instructional and curricular alignment can negatively affect student performance. For example, during the elementary and middle school years, students may experience problem-centered curricula resources in which sense making is central to students' learning but now find themselves in traditional high school classes where sense making is replaced with teacher-directed procedural instruction.

Challenge: Content

Students who have done well in middle school mathematics may still face challenges during the transition to high school. For many students across the country, algebra 1 is the first high school mathematics course they will take. Many schools offer an integrated mathematics sequence instead of traditionally organized courses such as algebra 1, geometry, and algebra 2. Whether algebra 1 is taught as part of an integrated program or in the traditional sequence of single-course emphases, similar issues can arise.

Regardless of whether algebra 1 is taught in high school or in middle school, how well students do in this class can have a lasting effect on the rest of their high school mathematics study. Students' success in algebra 1 may hinge on the quality and quantity of prior experiences in which they learn to reason about patterns, relationships, and representations. Also, it matters whether those prior experiences were closely aligned to the ways that algebra is taught and assessed in high school.

The ways in which teachers at various grade levels perceive algebra may also differ. For example, if algebra is perceived and taught as a way of solving equations (i.e., "find x," where x stands for a number), then students may be inadequately prepared for a functions-based algebra course. The CCSSM emphasizes the importance of proportional reasoning for success in algebra. If students are to be successful in high school mathematics, it is critical that they have ample opportunity to develop proportional reasoning skills across the middle grades. Facility with proportional reasoning must go beyond learning to use cross multiplication to solve for x in proportions such as

$$\frac{9}{15} = \frac{x}{10}.$$

Opportunities to reason proportionally can be fostered across mathematical concepts or strands and across representations. Both probability and measurement offer fertile ground for exploring proportional reasoning. Also, greater understanding of proportions can be developed when students are encouraged to reason about relationships among the numbers. For the equation

$$\frac{9}{15} = \frac{x}{10},$$

students who use mathematics flexibly may notice that the first ratio can be represented differently to work with it for different purposes. Instead of employing cross multiplication, the student can reason using scaling (or scale factors) and find that

$$\frac{9}{15} = \frac{3}{5} \text{ and } \frac{3}{5} = \frac{x}{10},$$

and then easily conclude that

$$\frac{9}{15} = \frac{6}{10}$$

to determine that $x = 6$. Students' ability to *re*-represent numerical values is essential to later success in the more formal equation solving that is required in algebra and throughout high school.

Challenge: Psychological and Social Factors

The challenges of the transition to high school extend beyond issues of instructional and curricular alignment and of the mathematics content itself. The National Mathematics Advisory Panel (2008) surveyed several hundred teachers to better understand their experiences teaching algebra. The panel's final report stated that 62 percent of the teachers surveyed reported "working with unmotivated students" as the single most challenging aspect of successfully teaching algebra 1. If we want to engage students in meaningful mathematics, it is critical that we do not underestimate what it takes to motivate them to be successful in school.

One important component of student motivation has to do with students' self-beliefs—for example, whether students believe that their intelligence is *fixed* or *malleable*. Students who hold the latter view tend to pursue challenges and are more resilient in the face of setbacks. Those students who view intelligence as being fixed tend to pursue easy successes and are more vulnerable in the face of setbacks. Pajares and Schunk (2002, p. 18) assert,

> It may even be reasonably argued that teachers should pay as much attention to students' self-beliefs as to actual competence, for it is the belief that may more accurately predict students' motivation and future academic choices.

Peer acceptance and influence are of increasing importance in determining students' self-efficacy during early adolescence. Students' sense of belonging in the mathematics classroom can significantly affect whether—and the degree to which—they choose to engage in their own learning. Yet students' sense of belonging in the academic community is complicated by the fact that within the culture of the typical U.S. high school, a student's immediate social concerns often have far more motivational power than that student's academic concerns. In many students' eyes, activity in these two dimensions (academic and social) is not only separate but also can actually be contradictory—that is, students may see success in the academic arena going hand in hand with failure in the social arena.

Easing Students' Transition to High School

We can guide students to help overcome these transition-to-high-school challenges in several ways. We can—

1. improve instructional alignment through collaboration between middle and high school teachers;
2. address mathematics content challenges using structured professional learning communities; and
3. shape classroom culture to address some of the social and psychological factors that students face.

Suggested Practice: Greater Alignment through Collaboration

What can teachers and other educators do to promote a greater degree of success for their students making the transition to high school? A first step can be to collaborate on issues that relate to this transition. Too often, teachers work in isolation, each trying to solve problems alone without a systematic and coordinated effort. DuFour and his colleagues illuminate a growing consensus among education leaders that this isolation must end. They advocate the use of collaboration as a communication, articulation, and problem-solving tool (DuFour, Eaker, and DuFour 2005).

However, this kind of collaborative, in-depth work, especially when it is used to deal with a transition involving multiple schools, requires genuine cross-campus collaboration among mathematics faculty, not just conversations among district leaders.

Middle and high school teachers need opportunities to discuss transition issues, collaboratively study and understand their state mathematics standards, and work together to better prepare students for success in learning to use increasingly abstract notations and representations. This collaboration can occur only when colleagues work together to develop common understandings about carefully aligned and developmentally appropriate resources and effective instructional strategies.

In this collaboration, teachers can clarify what content each is responsible for teaching, how that content is taught, and how it could be assessed. Eighth-grade teachers should be aware of exactly how algebraic reasoning was developed and promoted by the seventh-grade teachers. Their teaching should then connect to those understandings and experiences and consider exactly how far to extend them. Likewise, high school teachers need to be able to pick up where the eighth-grade teachers left off so that their teaching can confidently connect to what students have already learned. Such boundary-crossing means that teachers need structures that allow these collaborations to occur.

At the student level, the transition from middle school to high school coursework should feel seamless and well connected. Students should not view high school mathematics as something completely new. It should instead reflect a natural progression of concepts and ideas for which students have been prepared.

Suggested Practice: Learning Communities Focused on Instruction and Ongoing Assessment

A natural extension of collaborative work is to develop more structured professional learning communities to help students transition from middle to high school mathematics. The idea of forming such groups among teachers is certainly not new; done well, however, the process involves far more than simply gathering educators together. An effective professional learning community should address questions such as these:

1. How should the community structure work around a particular problem?
2. What role does each participant play?
3. What evidence will be used to measure progress in student outcomes?

Several good models for organizing effective learning communities exist. Of particular note, Driscoll and colleagues at the Education Development Center (EDC) in Massachusetts have been developing resources to help teachers collaboratively study the effective teaching and learning of mathematics. The core belief guiding the EDC work is that good mathematics instruction begins with understanding how mathematics is learned.

Their work also emphasizes the use of classroom artifacts (e.g., student work samples, transcripts and video of classroom interactions) to provide common ground for teacher discussion and promote a greater degree of content knowledge in both mathematics and pedagogy. For example, *Fostering Algebraic Thinking: A Guide for Teachers, Grades 6-10* (Driscoll 1999) and its companion book *The Fostering Algebraic Thinking Toolkit: A Guide for Staff Development* (Driscoll et al. 2001) clarify how algebraic "habits of mind" (see chapter 9 in this volume) can be promoted across grades 6–10. These resources advocate the use of problem-solving scenarios that promote algebraic reasoning as well as offer strategies for finding algebra in unlikely places so that students will not view it as something separate from the rest of mathematics. The notion of teachers forming professional learning communities to address instruction and curriculum alignment holds tremendous promise (Paek 2008).

Suggested Practice: Shaping Classroom Culture

Several districts and schools have successfully taken on the problems that go beyond transition, such as students' motivation and self-beliefs. Some have undertaken modest interventions designed to more actively engage students in doing mathematics and create a stronger academic culture that supports students' cognitive, social, and motivational development.

For example, the Academic Youth Development (AYD) program seeks to create a positive culture for learning high school mathematics by shaping the existing classroom culture by using ideas from cognitive science, neuroscience, and educational psychology. (The AYD program was created by the Charles A. Dana Center at the University of Texas at Austin in collaboration with Agile Mind.) This program engages students in discussions and activities that teach them how people learn, how to overcome obstacles to learning, and how to create a community of learners through mutual accountability.

These youth development or psychological strategies are taught in parallel to a component focusing on ratio and proportional reasoning, multiple representations of relationships, and problem solving. AYD supports incoming ninth-grade algebra students by engaging them in a summer bridge class before their algebra course, which focuses on all these elements and then reinforces them by strategically timed school-year experiences. The AYD program aims to teach students how to think and reason mathematically. It also works to shape the ways that students think about themselves as learners, develop their commitment to high achievement, and create a set of social supports that sustain their responsible and productive engagement in challenging courses.

One of the most important lessons learned from this program is that students' mathematical learning and high school success can improve greatly when they realize that their intelligence is *malleable*, not *fixed*. This finding suggests that we can ease students' transition from middle school to high school math by incorporating into the teaching of math these various potent ideas from social psychology:

1. Effective effort
2. Attribution of effort
3. The significance of interpersonal skills
4. A sense of belonging
5. Motivation

It is powerful to draw on neuroscience to show students how their brains actually change as they learn new things. As mathematics teachers, we may limit our effectiveness if we focus only on mathematics without taking into account the psychological factors that are so critical to student success—especially for students at such a tumultuous stage of their lives.

Next Steps

If we want to improve students' transition from middle school to high school mathematics, we need to meet with colleagues and consider ways to work together on this problem—both on and across campuses. The alignment of both instruction and curricula is critical. But what we teach and how we teach it are the most important factors within the control of teachers. Toward that end, we conclude with the following recommendations:

- Commit to forming a professional learning community around the issue of transitioning between schools and approaches to mathematics, and request the support needed to progress toward improvement.

- As a learning community, determine how you will know that you are making progress, and hold yourselves accountable.
- Remember to consider factors beyond mathematics, such as social and psychological issues, and work toward building a school climate that promotes positive and productive student self-beliefs.

Students always face a challenge as they shift from one level to the next. But we, as teachers, can greatly increase the likelihood that students will successfully meet this challenge and experience success in high school mathematics.

References

Charles A. Dana Center. "Academic Youth Development." Austin, Tex.: University of Texas at Austin, 2011. http//:www.utdanacenter.org/academicyouth.

Common Core State Standards Initiative. *Common Core State Standards for Mathematics*. Washington, D.C.: National Governors Association Center for Best Practices and the Council of Chief State School Officers, 2010. http://www.corestandards.org.

Driscoll, Mark. *Fostering Algebraic Thinking: A Guide for Teachers, Grades 6–10*. Portsmouth, N.H.: Heinemann, 1999.

Driscoll, Mark, Judith Zawojewski, Andrea Humez, Johannah Nikula, Lynn T. Goldsmith, and James Hammerman. *The Fostering Algebraic Thinking Toolkit: A Guide for Staff Development*. Portsmouth, N.H.: Heinemann, 2001.

DuFour, Richard, Robert Eaker, and Rebecca DuFour, eds. *On Common Ground: The Power of Professional Learning Communities*. Bloomington, Ind.: Solution Tree, 2005.

Eccles, Jacquelynne S., Sarah Lord, and Carol Midgley. "What Are We Doing to Early Adolescents? The Impact of Educational Contexts on Early Adolescents." *American Journal of Education* 99 (August 1991): 521–42.

National Mathematics Advisory Panel. *Foundation for Success: The Final Report of the National Mathematics Advisory Panel*. Washington, D.C.: U.S. Department of Education, 2008.

Paek, Pamela L. "Practices Worthy of Attention: Local Innovations Strengthening Secondary Mathematics." Austin, Tex.: Charles A. Dana Center at the University of Texas at Austin, 2008. http://www.utdanacenter.org/pwoa/downloads/columbus.pdf. 2008.

Pajares, Frank, and Dale Schunk. "Self and Self-Belief in Psychology and Education: A Historical Perspective." In *Improving Academic Achievement: Impact of Psychological Factors on Education*, edited by Joshua Aronson, pp. 3–21. San Diego, Calif.: Academic Press, 2002.

Chapter 19

High School to Postsecondary Education: Challenges of Transition

Susan Hudson Hull
Cathy L. Seeley

The Common Core State Standards for Mathematics (CCSSM) focus on the mathematical knowledge, skills, and practices that high school students need for success in college and careers (Common Core State Standards Initiative 2010). But beyond attention to the content expectations, how can we help prepare our high school students to make the critical transition to what comes after they graduate? We cannot wait until students are seniors to inspire them to consider postsecondary education or training and at the same time prepare them to succeed when they get there.

It is increasingly important for students to continue their education beyond high school, not only for their future—both opportunities and earnings—but also for our country's future (Achieve, Inc. 2008; Education Trust 2003). Recognizing this, President Obama, in his first address to Congress in 2009, set a goal of postsecondary education for every American (Associated Press 2009).

The good news is that a recent study found that 90 percent of low-income students intend to go to college after graduation from high school, with little gap between white and minority students' college-going intentions. The bad news is that the same study shows that only half actually enroll (Associated Press 2009).

The Problem

Unfortunately, adults often do not have the same expectations for students as students have for themselves. A 2000 study by the Education Trust (2003) reported that 79 percent of students surveyed planned to go to college, but only 51 percent of their teachers and 68 percent of their parents expected these same students to pursue higher education (p. 21).

Often, high school graduates may not be prepared for college or the workforce, even if they have passed the required mathematics courses and exit tests for graduation. It is not enough for students to take first- and second-year algebra and geometry or their integrated course equivalents. (Some schools offer integrated high school mathematics courses in which the traditional sequence of first-year algebra, geometry, and second-year algebra are replaced with three years of integrated mathematics. Throughout this chapter, we use second-year algebra to include the third course in this alternative organizational scheme.) Many studies show that taking an advanced mathematics course beyond second-year algebra makes a tremendous difference in student preparedness, as does taking mathematics each year in high school (Adelman 2006).

Teachers' expectations matter most. In one survey, 76 percent of the college students who responded reported that they were only somewhat challenged by their high school courses or that it was pretty easy to slide by in their courses (Achieve, Inc. 2005). Sixty-five percent of those who were enrolled in college and 77 percent of those not enrolled in college reported that, in hindsight, if high school courses had been more demanding, they would have applied themselves more diligently, worked harder, and taken more challenging courses.

Often students do not have information about what colleges require—for entrance or for placement. About 50 percent of students entering college, even those who have taken second-year algebra and precalculus, need some remediation (Venezia, Kirst, and Antonio 2003), which means they are placed into non-credit-bearing courses so that they can relearn (or learn) high school mathematics. As one student said, "They showed me how to fill out a McDonald's application in my Life Skills class. I think that they should have at least taught me how to fill out a college application or told me what the college requirements are" (Rivera and Alcalá 2005).

Then there are those students who say, "But I'm not going to college, so I don't need math." Yet 80 percent of the fastest growing jobs require postsecondary education or training. Career and technical guides for students indicate that a wide range of careers in health care, manufacturing, finance, construction, and others recommend a year of mathematics beyond the level of second-year algebra (Achieve, Inc. 2008). There is growing consensus that if a student wants at least a "family wage" job, then preparation to achieve college readiness and preparation to achieve work readiness are essentially the same (ACT 2006).

Bridging the Gap

The worlds of high school and postsecondary education, especially with regard to mathematics, may seem light-years apart. How often do high school teachers and higher education faculty talk with each other about content knowledge or their expectations of students? How often do they communicate about articulation between their programs? How much do high school teachers know about the college entrance exams or placement tests used by the institutions where most of their students apply?

Many state standards for high school mathematics are being revisited to include college-readiness standards, yet the understanding of what these college-readiness standards mean often varies, especially across the high school–postsecondary school divide. How well do districts and teachers ensure that the second-year algebra taught in high school is the algebra needed for postsecondary success?

In *College Knowledge* (2008), Conley recommends that high school and college faculty work together to develop a common understanding of what students need to know, to create tasks that exemplify these expectations, and to decide jointly on the quality of student work that will indicate appropriate depth of knowledge. He suggests that students as well as teachers and administrators participate in these conversations. Adelman (2006) suggests that students, teachers, and their parents need to see samples of first-year college mathematics assignments and exams to get a better idea of what students will be expected to do.

Increasing High School Requirements for All

With the increased emphasis on college and work readiness, many states are strengthening mathematics requirements for graduation. Ten years ago, many states did not require even first-year algebra for all students, but now second-year algebra or its equivalent is an increasingly common requirement. However, each state sets its own standards—and, therefore, definitions—for courses, so a credit in second-year algebra may mean different things in different states. In fact, a credit in second-year algebra in one classroom might mean something different from a credit in second-year algebra in a classroom across the hall in the same school. To try to ensure some accountability and consistency, several states are beginning to require end-of-course tests in key high school mathematics courses.

Although there is increasing agreement about the need for four years of high school mathematics, few argue that this means that all students should take a precalculus course. Calculus is not required for many college degrees outside science, technology, engineering, and mathematics (STEM) majors, but statistics is steadily becoming more essential. Thus, a fourth-year high school mathematics course that stresses statistics and probability as well as data and modeling could be of significant value in helping students bridge the transition between high school and college or the world of work.

What Messages Can We Give Our High School Students?

Classroom teachers can work at a personal level to help students shift their thinking about preparing for their next steps after high school by discussing the following ideas with them:

- *Take mathematics every year.* Taking mathematics every year in high school makes a huge difference in postsecondary success. Advise students that choosing courses that challenge them and prepare them for further study or training is critical. Mathematics through the second-year algebra level or its equivalent is essential for all students, whether for college or for work, and each course beyond this level dramatically increases the likelihood of future success. Counsel students to make sure that they are enrolled in good mathematics courses every year of high school and encourage them to start mathematics in their first semester of college.

- *Prepare yourself for more than just getting into college.* Let students know that placement tests may dictate whether they can take college-credit courses or whether they will be placed in remedial courses that carry no credit. Encourage them to investigate placement tests offered at their choices of college and to be prepared for these tests. Many colleges have online information that can help.

- *Understand the role of calculators.* Remind students that calculators are incredible tools for learning mathematics and solving problems, but they are not crutches to absolve them of responsibility for algebraic fluency. Postsecondary education faculty often set their own calculator policies; students must understand mathematics in multiple ways so that they are ready for whatever policy arises.

- *Know how to read for understanding.* Being able to understand the academic vocabulary of mathematics and to read and understand mathematics textbooks are essential skills to learn before leaving high school.

- *Form a study group.* Many successful college students rely on study groups to maximize their learning. Understanding and learning are often best attained in a group setting, where students can ask questions of one another, verbalize their thinking, and push one another to higher levels. They do not have to learn college mathematics all on their own.

- *Accept that mathematics can be difficult.* Learning mathematics can be, at various times, challenging for anyone. It is unrealistic to think that unless mathematics comes easily, one will not be able to do it or that one is not a "math person." The most successful students are those who are persistent when facing challenges, who have multiple options when they need to back up and start over, and who are willing to seek help when they are truly stuck.

What Can We Do to Ease Students' Transition?

As teachers, we can also challenge ourselves to prepare students better for what lies ahead:

- *Expect a lot from students.* Communicate to students the expectation that each of them will participate in postsecondary learning (whether in two- or four-year colleges), in certification programs, or by entering a "family wage" job that is likely to involve continued training. We must help students keep their options open; high school students often do not realize that the choices they make now will have ramifications and that later they may want to make other choices. They need four years of high school-level mathematics, including the content of first-year algebra, geometry, and second-year algebra as well as a strong mathematics course beyond second-year algebra. All students may not need a precalculus course, but all need a strong background in data and statistics, modeling, algebraic and geometric reasoning, and numeracy (as well as problem solving and communicating, representing, and connecting their understanding) from which to branch into their postsecondary course of study.

- *Build solid relationships with students.* Students need adults who care about them and hold them to high standards. Students often approach their teachers for college advice. When they turn to you, let them know what you expect regarding their mathematics preparation. Keep in mind the studies in which students report that they would have taken more rigorous mathematics courses or studied more in high school had they been expected to do so.

- *Talk to higher education faculty in your area.* Find out what they expect from first-year students entering their classes—the skills they want to see, their calculator policies, the support they provide to students, and the quality of work they expect. Consider the potential benefits to all participants if high school and higher education faculty were to agree on a set of common assessment items for core content, administer the assessments in equivalent high school and college courses, and then evaluate them together to decide "how good is good enough."

- *Be aware of what is happening at the local, state, and national level.* College and career readiness are timely topics at all levels. If your state has college- and work-readiness standards that are consistent with, or supplement, the CCSSM, know what they are and help your students know what they are. Look for external validation of your course, assessments, and criteria for judging "how good is good enough" for your students.
- *Know and communicate the importance of preparing for college placement tests.* These tests often relegate students to remedial courses to relearn mathematics you know they know. Placement tests vary from campus to campus. Find out what placement tests your area colleges use and share this information with your students. Consider hosting review sessions before the test.

As in everything else we do in teaching, what we expect of students and how we support them in meeting these expectations can make all the difference in what they learn. If students are to successfully make the transition from high school to postsecondary education or the workforce, they need to think ahead, make informed choices, and work hard. Our responsibility as mathematics teachers is to support them in each of these pursuits so that all students have the mathematics they need for a productive and rewarding future.

REFERENCES

Achieve, Inc. *Rising to the Challenge: Are High School Graduates Prepared for College and Work?* Washington, D.C.: Achieve, Inc., 2005. http://www.achieve.org/RisingtotheChallenge.

———. *Math Works.* Washington, D.C.: Achieve, Inc. 2008. http://www.achieve.org/mathworks.

ACT, Inc. *Ready for College and Ready for Work: Same or Different?* Iowa City, Iowa: ACT, 2006.

Adelman, Clifford. *The Toolbox Revisited: Paths to Degree Completion from High School through College.* Washington, D.C.: U.S. Department of Education, 2006. http://www.ed.gov/rschstat/research/pubs/toolboxrevisit/toolbox.pdf.

Associated Press. "College for All: Is Obama's Goal Attainable?" February 28, 2009. http://www.msnbc.msn.com/id/29445201.

Common Core State Standards Initiative. *Common Core State Standards for Mathematics.* Washington, D.C.: National Governors Association Center for Best Practices and the Council of Chief State School Officers, 2010. http://www.corestandards.org.

Conley, David T. *College Knowledge: What It Really Takes for Students to Succeed and What We Can Do to Get Them Ready.* San Francisco: Jossey-Bass, 2008.

Education Trust. "A New Core Curriculum for All: Aiming High for Other People's Children." *Thinking K–16* 7 (Winter 2003): 1–2. http://www.edtrust.org/dc/publication/a-new-core-curriculum-for-all-aiming-high-for-other-peoples-children-0.

Rivera, Selene, and Christian Alcalá. "Coalition Demands Access to Higher Education." March 24, 2005. www.innercitystruggle.org/story.php?story=129&print=1.

Venezia, Andrea, Michael Kirst, and Anthony Antonio. "Betraying the College Dream: How Disconnected K–12 and Postsecondary Education Systems Undermine Student Aspirations." Final report, Stanford University's Bridge Project, 2003, p. 8.